HAZARDOUS MATERIALS AIR MONITORING AND DETECTION DEVICES

HAZARDOUS MATERIALS AIR MONITORING AND DETECTION DEVICES

CHRIS HAWLEY

DELMAR
THOMSON LEARNING™

Australia Canada Mexico Singapore Spain United Kingdom United States

DELMAR
THOMSON LEARNING™

NOTICE TO THE READER

Publisher does not warrant or guarantee any of the products described herein or perform any independent analysis in connection with any of the product information contained herein. Publisher does not assume, and expressly disclaims, any obligation to obtain and include information other than that provided to it by the manufacturer.

The reader is expressly warned to consider and adopt all safety precautions that might be indicated by the activities herein and to avoid all potential hazards. By following the instructions contained herein, the reader willingly assumes all risks in connection with such instructions.

The publisher makes no representation or warranties of any kind, including but not limited to, the warranties of fitness for particular purpose or merchantability, nor are any such representations implied with respect to the material set forth herein, and the publisher takes no responsibility with respect to such material. The publisher shall not be liable for any special, consequential, or exemplary damages resulting, in whole or part, from the readers' use of, or reliance upon, this material.

Delmar Staff
Business Unit Director: Alar Elken
Executive Editor: Sandy Clark
Acquisitions Editor: Mark Huth
Editorial Assistant: Dawn Daugherty
Executive Marketing Manager: Maura Theriault
Marketing Coordinator: Brian McGrath
Executive Production Manager: Mary Ellen Black
Project Editor: Barbara L. Diaz
Art & Design Coordinator: Rachel Baker

Printed in Canada

1 2 3 4 XXX 03 02 01 00

For more information, contact Delmar at 3 Columbia Circle, PO Box 15015, Albany, New York 12212-5015; or find us on the World Wide Web at http://www.delmar.com

Library of Congress Cataloging-in-Publication Data

Hawley, Chris.
Hazardous materials air monitoring and detection devices / Chris Hawley.
p. cm.
Includes bibliographical references and index.
ISBN 0-7668-0727-4 (alk. paper)
1. Hazardous substances. 2. Air—Pollution—Measurement. 3. Air sampling apparatus. I. Title.

T55.3.H3 H377 2001
628.5′3′0287—dc21 2001017083

Asia (including India):
Thomson Learning
60 Albert Street, #15-01
Albert Complex
Singapore 189969
Tel 65 336-6411
Fax 65 336-7411

Australia/New Zealand:
Nelson
102 Dodds Street
South Melbourne, Victoria 3205
Australia
Tel 61 (0)3 9685-4111
Fax 61 (0)3 9685-4199

Latin America:
Thomson Learning
Seneca 53
Colonia Polanco
11560 Mexico D. F. Mexico
Tel (525) 281-2906
Fax (525) 281-2656

Canada:
Nelson
1120 Birchmount Road
Toronto, Ontario
Canada M1K 5G4
Tel (416) 752-9100
Fax (416) 752-8102

UK/Europe/Middle East:
Thomson Learning
Berkshire House
168-173 High Holborn
London WC1V 7AA
United Kingdom
Tel 44 (0)171 497-1422
Fax 44 (0)171 497-1426

Business Press
Berkshire House
168-173 High Holborn
London WC1V 7AA
United Kingdom
Tel 44 (0)171 497-1422
Fax 44 (0)171 497-1426

Spain:
Paraninfo
Calle Magallanes 25
28015 Madrid
España
Tel 34 (0)91 446-3350
Fax 34 (0)91 445-6218

Distribution Services:
ITPS
Cheriton House
North Way
Andover,
Hampshire SP10 5BE
United Kingdom
Tel 44 (0)1264 34-2960
Fax 44 (0)1264 34-2759

International Headquarters
Thomson Learning
International Division
290 Harbor Drive, 2nd Floor
Stamford, CT 06902-7477
USA
Tel (203) 969-8700
Fax (203) 969-8751

CONTENTS

PREFACE

This text covers a wide variety of air monitoring devices, including some basic discussion of their detection technologies. Part of this text provides information on how to interpret the readings these air monitoring devices provide. It is this interpretation that is the most difficult to present. The backbone to this interpretation is risk-based response (RBR), which may be a new concept to some readers. This system is very important to the responders' health and safety, and it is necessary to provide the background on the development of the system. RBR theory originated from several sources. Some concepts were developed by a good friend, Frank Docimo, who provided a truly simplified, street-smart approach to HAZMAT response. He laid the groundwork that was built up into RBR by Buzz Melton, former HAZMAT Chief for the Baltimore City Fire Department and now an environmental chemist with FMC Baltimore. We were developing a technician and technician refresher program, and the idea for RBR was born out of that development. We tried many ideas, and narrowed them down to a concept that is close to what is used today. We developed a dog and pony show presenting this concept, which was well received nationwide. I would be remiss if I did not thank our road crew, who made us look good during the worst of times. Although we no longer teach this program together, we still teach many other programs together and collaborate on other projects. From this program I modified some of the concepts into what we now call RBR. Our HAZMAT and many other HAZMAT teams across the country use this concept regularly and it is part of their daily response profile. Changing to RBR use was slow, taking several years, but the end result was worth it. Our training evolved to a level that we started using live chemicals during exercises. On a regular basis for training scenarios, we use oleum (concentrated fuming sulfuric acid) and probably one of the worst acids responders would ever face. The benefit is that once having trained with the worst material, any other spill of a corrosive cannot be any worse than the one used for training. This training, especially with the oleum, was the final step in our transformation to using RBR as an everyday response profile. Fortunately you can benefit from these years of hard work, training, and experience. One caution though—used in its pure form, RBR can be a dramatic change, and sometimes requires thinking outside the conventional HAZMAT box. When dealing with rescues, public endangerment, and terrorism there currently is no better system to keep responders safe and give them the ability to quickly handle any event.

ACKNOWLEDGMENTS

There is no way I could thank all those who have helped me throughout my career. I could provide a listing of names, but invariably I would miss some people. With that said, here is a feeble attempt at a list. The first people that I wish to thank are my sons, Timothy and Mathew, who are a constant source of inspiration and fun. Several friends have been key in helping me along with my career and this text, and they are John Eversole, Steve Patrick, Greg Noll, Mike Hildebrand, Mike Callan, Luther Smith, Gary Warren, and my road partner PJ Cusic. Robert Swann, one of the most knowledgeable responders I know from the Maryland Department of Environment (MDE), was an enormous help in the development of this text and my other texts, and always has an answer to my questions. The other responders from MDE deserve my thanks as well for all of their assistance. Armando (Toby) Bevelaqua helped me tremendously when I got stuck and got me through some serious writer's block. The flow charts in Chapter 10 were finished through his assistance and advice. I would be remiss if I did not mention my saga friends who live my life via the internet, but I specifically want to thank Stephanie, Jeanne, Trish, and Lisa who have been with me through thick and thin, and have made more than one moment fun and exciting. The personnel assigned to the HAZMAT team and those specifically at Station 14 deserve a lot of credit for their hard work and patience. The crew at Station 3 B deserves thanks for putting up with me during my stint as a shift instructor. As always make sure you have fun (rule # 1) and don't get killed (rule # 2), always ask questions and never stop learning.

The author and Delmar would like to thank the following persons for their review and assistance with this text:

Robert Royall,
HAZMAT Captain,
Houston Fire Department

Glen Rudner,
Hazardous Materials Officer,
Virginia Department of Emergency Management

Dan Law,
Consultant,
Mariposa, California

Gene Reynolds,
Environmental Chemist,
FMC Baltimore

Chris Wrenn,
Marketing Director,
RAE Systems

Bernie Edmonson,
Special Operations Coordinator,
Ft. Meade Fire Department, Maryland

Larry J. Broockerd,
LJB & Associates,
Merriam, KS

Ronald C. Thomas, Jr.,
Florida State Fire College,
Ocala, FL

ABOUT THE AUTHOR

Christopher Hawley is a Fire Specialist with the Baltimore County Fire Department, currently assigned as the HAZMAT Coordinator, a position he has held for nine years. He has been a HAZMAT responder for more than fourteen years. Chris has twenty-one years experience in the fire service and has been with the Baltimore County Fire Department for twelve years, with assignments in fire suppression, fire and rescue training, and hazardous materials, including an assignment as a shift instructor. Prior to that he was a Hazardous Response Specialist with the City of Durham, North Carolina, Fire Department. He has served as a volunteer in Loganville, Monroeville #4, and Richland Township (Johnstown) Pennsylvania, and has served in a variety of ranks including chief.

Chris has designed innovative programs in hazardous materials and has assisted in the development of other training programs, many for the Maryland Fire and Rescue Institute (MFRI). Chris is an adjunct instructor to the National Fire Academy and has served on development committees. He currently serves as a codeveloper for Emergency Response to Terrorism: Tactical Considerations HAZMAT course and as technical advisor and reviewer of the other terrorism programs. Chris has presented at numerous, local, national, and international conferences. He serves on a variety of pivotal committees and groups at the local, state, and federal levels. He works with local, state, and federal committees and task forces related to hazardous materials, safety, and terrorism.

Chris is the owner of FBN Training, which provides a wide variety of emergency response training, including hazardous materials, confined space, technical rescue, and emergency medical services, as well as consulting services to emergency services and private industry. As a subcontractor to Community Research Associates (CRA), he cowrote the FBI Hazardous Materials Operations training program and through CRA regularly serves as a course manager for these programs. Also through CRA he served as a technical consultant for the hazardous materials Technician level program, and serves as one of three course managers for that training program. He has served as course manager for a hazardous materials training program for specialized assets of the FBI nationwide. Through FBN Chris is a technical consultant to a number of air monitor manufacturers, including those who manufacture warfare and terrorism agent detectors. Also through FBN Chris continues to serve as a technical consultant to a Department of Defense contractor specifically hired to provide information related to a fire department's response to a terrorist incident.

ROLE OF AIR MONITORING IN HAZMAT RESPONSE

INTRODUCTION

One of the most important parts of the response to chemical incidents, air monitoring, is one of the primary mechanisms for keeping response teams alive and healthy. Hazardous materials (HAZMAT) teams are unique entities within the fire service and provide a valuable service to firefighters and the community. In some cases, HAZMAT teams take risks for the protection of life and property, which sometimes extends beyond the normal building, home, or vehicle to include water, air, land, trees, and the general environment. HAZMAT teams are sometimes requested because other firefighters have encountered a situation that is outside their scope of expertise, that has endangered or injured them, or that presents a unique risk they are uncomfortable dealing with. Why this focus on the unique challenges to a HAZMAT team? Air monitoring is a major factor in keeping the HAZMAT team members safe and should factor into every response.

SAFETY Air monitoring protects responders from an unidentified or unexpected situation, and items such as heat stress, slip, trip, and fall hazards may be greater dangers.

Other forms of protection, such as a Level A suit, may be excessively protective and could create a more harmful situation than it is intended to protect against. Air monitoring protects responders from an unidentified or unexpected situation, and the risks related to the use of Level A suits such as heat stress or slip, trip, and fall hazards may be greater dangers. In emergency response, it is necessary to identify the risk the material presents, make the situation safer, and then under less stressful conditions return the situation to normal. A common response-related question is what happens when a truckload of chemicals is mixed? Unless the materials react with each other, there is little concern if effective air monitoring is performed. Usually any violent reactions will occur prior to arrival of the responders, and if other reactions are going to take place, there is usually sufficient warning. By following a **risk-based response** (RBR) philosophy, the exact identity of the material does not matter. The RBR focuses on the immediate fire, corrosive, and/or toxic hazard, hazards that can be protected against by some form of protective clothing. The use of air monitors allows you to do the job of saving lives and property as you were sworn to do. This lifesaving ability is demonstrated in the following case study Drums and Food on I-95 and in Figure 1-1. An air

CASE STUDY

Drums and Food on I-95—A truck carrying food rear ended another tractor trailer stopped in the middle lane of the highway. It is estimated that the food truck was traveling at 65 mph when it hit the stopped truck, which was carrying a mixed load consisting of mostly 55-gallon drums. The driver of the stopped truck was not hurt and was able to remove his shipping papers.

The driver of the food truck was alive but was pinned in the wreckage as can be seen in Figure 1-2. He was conscious and alert, and his initial vital signs were stable. The trucks were entangled, and the rear door to the truck carrying the drums was torn away. Drums had shifted onto the food truck and were lying over the front of that truck as shown in Figure 1-3. The first responders saw the placards on the first truck and requested a HAZMAT assignment. When they approached the truck to evaluate the driver, they saw the drums in a precarious position in the back of the truck. They already had turnout gear on but also donned self-contained breathing apparatus (SCBA). They obtained the shipping papers from the driver and consulted with the HAZMAT company, which was about twenty minutes away. The drums contained mostly flammable solvents and some combustible liquids. The first responders were advised to use air monitors, continue with full personal protective equipment (PPE), establish foam lines, and begin the rescue. Air monitoring, a vital form of protection for the rescuers, was continuous.

Upon arrival, the HAZMAT company met with the incident commander (IC) to evaluate the scene. They confirmed the monitoring being done by the first responders and began to evaluate the other parts of the load. It was determined that whenever the rescue companies would move a part of the dash of the truck, one of the leaking drums would increase its flow. The HAZMAT company secured the leak, moved the drum, and secured the remainder of the drums. They examined the rest of the load to make sure there were no other problems, monitored the atmosphere, and observed that most of the load had been deformed by the impact, but that there were no further leaks. Once the victim was removed and the rescue companies moved away, the HAZMAT team overpacked and pumped the contents of the other drums into other containers. The risk category was fire, and the PPE chosen was appropriate for the risk category. At no time were any flammable readings indicated during the rescue although some were encountered during the transfer operation. This example illustrates the various disciplines working together to rescue a victim in a hazardous situation, and air monitors providing a high level of protection for all.

Figure 1-1 Crews work to rescue the driver of the food truck on the right while HAZMAT crews remove drums of flammables from the other truck. Effective monitoring and risk-based response assisted in this rescue effort. (Photo courtesy of Baltimore County Fire Department.)

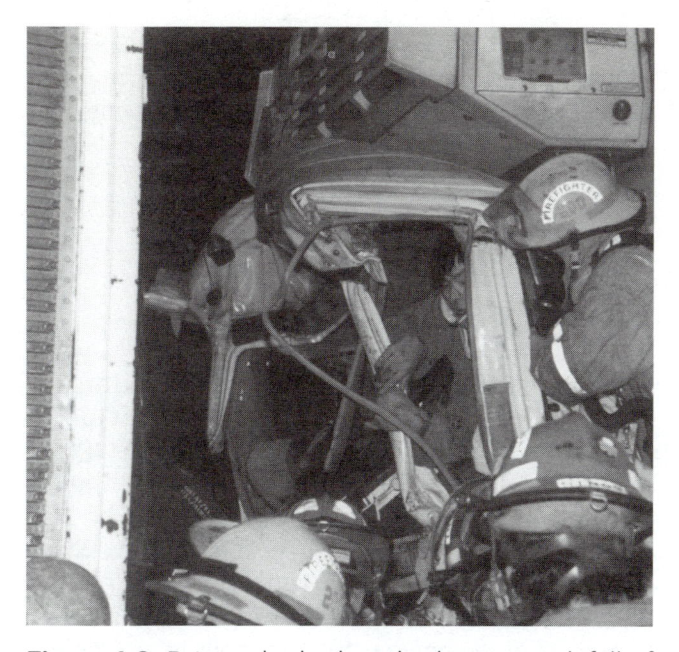

Figure 1-2 Driver who had crashed into a truck full of flammable drums is severely entangled in the wreckage. Rescue crews are operating in appropriate levels of PPE for the risk. (Photo courtesy of Baltimore County Fire Department.)

Figure 1-3 View of the other side of the truck, with the drums lying on the cab of the truck. The drums causing problems during the rescue were removed by HAZMAT crews. (Photo courtesy of Baltimore County Fire Department.)

monitor takes all of the gray out of HAZMAT response, and makes it black and white, provided you can interpret what it is trying to tell you. When the books tell responders to evacuate 7 miles, and the monitors tell responders 7 feet, which is more accurate? What reflects real-time, real-life situations? Which distance would make the incident less complicated?

Unfortunately air monitoring is not emphasized as it should be, nor is it commonly used in the way it was intended. For many years, air monitoring during emergency response was an afterthought, but it is now considered essential for personnel protection. A response team that is not adept at air monitoring is at an extreme disadvantage and can be placing itself and the public in harm's way. When searching for an unidentified chemical or dealing with multiple chemicals, you may find air monitoring and the new types of detectors overwhelming. In reality, although the use of air monitoring and sampling equipment requires practice, basic air monitoring has become very simple.

NOTE A response team that is not adept at air monitoring is at an extreme disadvantage and can be placing itself as well as the public in harm's way.

This text provides the responder with detailed information related to the role of air monitoring and air monitoring devices in an easy-to-read format. Most fire service responders have a reasonable understanding of **combustible gas indicators,** but that knowledge can be limited. Our society is becoming much more sophisticated and our

Figure 1-4 An essential device for HAZMAT crews, a photoionization detector detects the toxic risk portion of risk-based response.

customers expect more. Technology is available to HAZMAT teams, and it can be readily purchased. The cost of a photoionization detector (PID) as shown in Figure 1-4 is now less than $2,000, well within the reach of most departments. The cost is quickly recouped by the amount of information and the level of safety that can be obtained by using a PID. The following information presents some concepts on air monitoring, monitoring strategies, information on how the monitors work, and their uses. It is imperative that emergency responders understand what equipment is available and the limitations of that equipment.

In order to truly understand air monitoring we must be aware of the environment in which responders work every day. It is important to have a background in the many areas of hazardous materials response; however, it is impossible to know everything about all of the chemicals that exist. Currently about 50 million chemicals are listed by the **Chemical Abstract Service (CAS),** which is an overwhelming number. What are the most common responses for HAZMAT teams? What calls are run all of the time and what chemicals are being released every day? Once responders understand the fundamentals regarding monitoring, they can start learning the chemical and physical properties, as well as the hazards, associated with the ten most common chemicals, which are listed in Table 1-1. Once you learn the top ten, you will have covered most of the chemical families that responders need to deal with. Effective use of air monitoring will protect responders from the very unusual events that occasionally occur.

TABLE 1-1

Top Ten Chemicals Released
Ammonia
Sulfur dioxide
Chlorine
Hydrocholoric acid
Propane
Sodium hydroxide
Sulfuric acid
Gasoline
Flammable liquids
Combustible liquids

Note: This list was taken from a consolidated list of many different top ten lists provided by the Environmental Protection Agency (EPA) (Comprehensive Environmental Response Compensation and Liability Act (CERCLA), Emergency Incident Reporting and Notification System (EIRNS), and the National Response Center (NRC) data) report to Congress under the Clean Air Act Amendments (112 r) and Department of Transportation (DOT) (reported spills, fatalities, injuries).

REGULATIONS AND STANDARDS

Within the **Occupational Safety and Health Administration (OSHA) Hazardous Waste Operations and Emergency Response** (HAZWOPER) regulation (29 CFR 1910.120) there are not a lot of specific requirements for air monitoring. The **National Fire Protection Association (NFPA)** Standard 472 also has some requirements for air monitoring, but like the OSHA regulation, is fairly generic. The responder must be able to use detection equipment, and it must be used according to the manufacturer's recommendations. Both OSHA and the NFPA require you as a responder to be able to characterize an unidentified material. The HAZWOPER document includes more air monitoring information in the first part of the document, which discusses hazardous waste sites. The emergency response section (paragraph q) barely mentions air monitoring. It is the intention of 1910.120 to protect the worker, in this case the responders. OSHA requires the IC to identify and classify the hazards present on a site. Air monitoring is the primary way to fulfill this obligation. Once an effective monitoring strategy is developed, the IC only has to wait for the results. If the monitoring shows little or no readings, the incident may be of small significance and limited risk. But if the monitoring results in high readings, the risk increases, and the incident assumes a different level. If high readings are received, the results can assist with PPE decisions. The IC must also use air monitoring to determine the scope of the incident and any potential public protection options.

NOTE Both OSHA and the NFPA require you as a responder to be able to characterize an unidentified material.

NOTE OSHA requires the IC to identify and classify the hazards present on a site.

The NFPA has two documents that have an impact on air monitoring: NFPA 471, *Recommended Practice for Responding to Hazardous Materials*

Incidents and NFPA 472, *Standard for Professional Competence of Responders to Hazardous Materials Incidents*. Both provide some basic objectives related to air monitoring, including knowing the basic equipment and the need for operational checks and calibration. In the objectives for the HAZMAT Technician the NFPA requires a technician to be able to survey the HAZMAT incident to identify special containers involved, to identify or classify unidentified materials, and to verify the presence and concentrations of hazardous materials through the use of monitoring equipment. Another objective included by the NFPA is to have the technician interpret the data collected from monitoring equipment and to estimate the size of an endangered area. The technician must also know the steps for classifying unidentified materials. Being familiar with air monitoring is crucial to fulfilling these goals.

NOTE The NFPA requires a technician to be able to survey the HAZMAT incident to identify special containers involved, to identify or classify unidentified materials, and to verify the presence and concentrations of hazardous materials through the use of monitoring equipment.

GENERAL TERMINOLOGY

To be able to understand the remainder of this text, you must be familiar with some general air monitoring terminology. These terms apply to all air monitors and detection devices unless specifically noted. Later chapters detail specific information about each of the instruments. The suggested readings section provides texts that have further information on these and other terms.

Flammable Range

Flammable range comprises two ends, the **lower explosive limit (LEL)** and the **upper explosive limit (UEL),** which is the range in which there can be a fire/explosion when a flammable gas or vapor is mixed in proper proportion with air. The LEL is the lowest percentage of a flammable gas or vapor mixed with air that can be ignited. The UEL is the highest amount of a flammable gas or vapor mixed with air that can be ignited. Too little of a compound (too lean a mixture) and it cannot be ignited, and when there is too much of a compound (too rich a mixture) a fire is not possible either. Some texts call LEL the lower flammability limit (LFL), and the UEL the upper flammability limit (UFL). In order to have a fire or explosion the concentration of the flammable material must be within the **flammable range,** between the LEL and the UEL as shown in Figure 1-5. Combustible gas indicators (CGI) read up to the LEL point, and indicate fire or explosive situations. Some common materials and their flammable ranges are provided in Table 1-2.

NOTE The LEL is the lowest percentage of a flammable gas or vapor mixed with air that can be ignited. The UEL is the highest amount of a flammable gas or vapor mixed with air that can be ignited. Both require an ignition source.

CASE STUDY

One evening when I was a shift instructor on nightshift, I arrived at a firehouse to do some training. A couple of blocks from the station I passed a medic unit, which I assumed was headed toward a hospital. When I arrived in front of the station, the emergency medical services (EMS) district supervisor greeted me excitedly. He told me that I should have been there earlier as I had just missed a HAZMAT call. I had not heard any dispatch, so I was a little confused. He told me that there had been a gas leak in the basement and that it had exploded when the maintenance crew was relighting the pilot light on the hot water heater. He was concerned because while watching the operation he saw a big orange wall of flame pass by him and heard a woof sound. The only injuries were to the maintenance crew, but all were very lucky. The fire specialist on shift told me that, when told of the leak, the captain investigated it and stated that "It doesn't smell near the LEL [lower explosive limit]." There are a couple of issues here: First and foremost, the captain's nose missed the call, as there was an explosion, and second our noses are not calibrated to anything, so it is impossible for us to quantify a little, a lot, or a boatload. Many chemicals, particularly sulfur compounds such as those made to odorize natural gas, have a tendency to overwhelm the senses, and even though the material is still present, our noses shut off and no longer smell the gas. Responders should not use their noses to look for chemical odors, and should be in full PPE, which includes SCBA. This situation could have had disastrous results, with a major explosion and a potential building collapse.

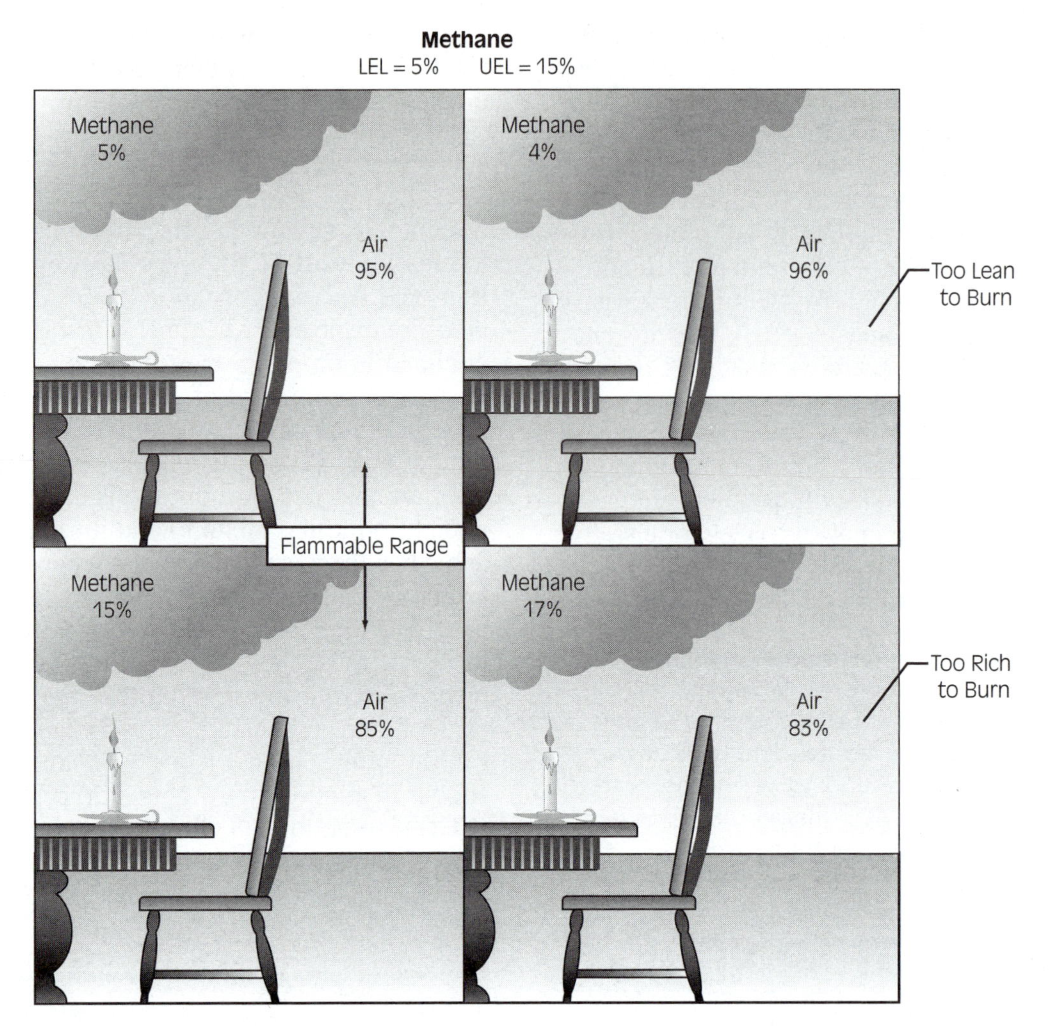

Figure 1-5 The flammable range is the range in which there can be a fire or explosion. Below the LEL or above the UEL there cannot be a fire.

TABLE 1-2

Flammable Ranges of Common Materials

Name	LEL (%)	UEL (%)
Acetylene	2.5	100
Acetone	2.5	12.8
Ammonia	15	28
Arsine	5.1	78
MEK (methyl ethyl ketone)	1.4 @ 200°F	11.4 @ 200°F
Carbon monoxide	12.5	74
Ethylene oxide	3	100
Gasoline	1.4	7.6
Hydrazine and UDMH (unsymmetrical dimethyl hydrazine)	2.9 (2%)	98 (95%)
Kerosene	0.7	5
Toluene	1.1	7.1
Hexane	1.1	7.5
Propane	2.1	9.5

Flash Point

Associated with the flammable range is the **flash point** of a material. The flash point is the temperature of a liquid that when heated, gives off sufficient vapors, that when mixed with air (into the flammable range) can be "flashed" (momentarily ignited) by an ignition source. There are several keys to this definition: (1) the temperature of the liquid, so that vapors can be given off; (2) the vapors mixing with air to reach the flammable range, and (3) an ignition source. The flash fire can really only be duplicated in the lab, but in the real world when something flashes, it generally continues to burn, thereby reaching its **fire point.** In the street the two temperatures are so close that a quick flash fire is not likely, but continued burning is most likely.

NOTE When a material boils, it is releasing itself into the air.

Boiling Point

The **boiling point** of a material is a clue as to how easily a detection device is going to pick it up. When a material boils, it is releasing itself into the air, changing the material's state of matter as shown in Figure 1-6. The lower the boiling point, the more rapidly the material moves to the gaseous (vapor) state under ambient conditions, and the easier it is picked up by a detection device. See Table 1-7 for a summary of the interrelationships of chemical and physical properties.

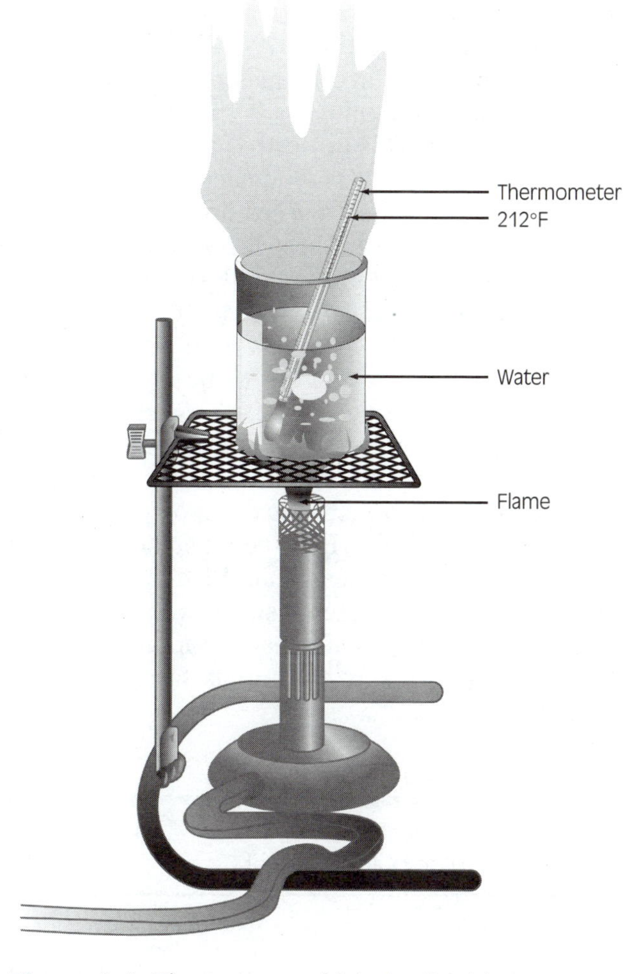

Figure 1-6 The point at which the liquid turns into a gas is called the boiling point.

Vapor Pressure

By far the most important physical property not only to HAZMAT but to air monitoring is **vapor pressure.** In Chapter 10 we discuss the importance of vapor pressure and risk-based response, and many of the issues provided here are further explained in that chapter. We can define vapor pressure in several ways, but the best is related to the amount of vapor that is emitted from a material. Both solids and liquids have vapor pressure, although it most often is associated with liquids.

NOTE The most important physical property, not only to HAZMAT but to air monitoring, is vapor pressure.

Another definition is the amount of force applied on a specific container by the vapors coming from a material at a given temperature. We measure vapor pressure by four common methods: millimeters of mercury (mm Hg), pounds per square inch (psi), atmospheres (atm), and millibars (mb). The values

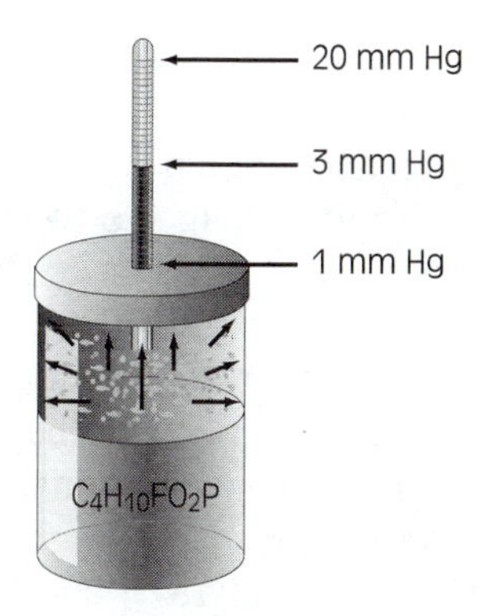

Figure 1-7 Vapor pressure is the force of the vapors pushing against the side of a container. In this drawing the vapors would push up the column of mercury, which would correspond to a reading of vapor pressure in millimeters of mercury (mm Hg).

of 760 mm Hg = 14.7 psi = 1 atm = 1,701 mb are all equal, and they represent normal atmospheric pressure. As shown in Figure 1-7, in order to determine the vapor pressure of a material, it is placed in a sealed container that has a thermometer-like device in the cap. This device has a column of mercury in a cylinder that is graduated in millimeters. The force of the vapors pushes up on this column of mercury, and the highest level attained is the vapor pressure at that given temperature. Most temperatures are recorded at 68°F, which is considered normal. We consider materials with a vapor pressure in excess of 40 mm Hg to be an inhalation hazard, or simply, a vapor hazard. When materials become vapors, they present more risk and are less easily controlled. We can compare vapor pressures with water, which has a vapor pressure of approximately 25 mm Hg. When a spilled material has a vapor pressure less than 25 mm Hg it will remain as a liquid longer than a given amount of water. If a material has a higher vapor pressure than water, then it will turn into a gas faster than the same amount of water.

NOTE The values of 760 mm Hg = 14.7 psi = 1 atm = 1,701 mb are all equal.

A third definition for vapor pressure is the most scientific: It is the force of the vapors as related to overcoming atmospheric pressure that is emitted from the surface of the material at a given temperature, overcoming the force that keeps the vapors in the material. Vapor pressure is the force that comes off of a material, and is related to another term, volatility. Often these terms are mistakenly used interchangeably. **Volatility** is the relative quantity of the vapors that come from the material versus a specific standard (normally) at a specific temperature. Volatility is provided in a milligram per meter cubed (mg/M^3) format, which can be converted to parts per million (ppm) using the formula found in Table 1-8. In most cases the relationship is consistent, as most high vapor pressure materials also have high volatility, but this does not hold true for all materials. For example, xylene has a low vapor pressure (9 mm Hg) but is a flammable liquid that evaporates fairly quickly, because its volatility is somewhat high (39, 012 mg/M^3) and because of its molecular structure. The chemical structure is a benzene ring and is highly unsaturated, which becomes a major factor in it being a flammable liquid. Table 1-3 provides some additional examples of chemical and physical properties.

SAFETY When materials become vapors, they present more risk and are less easily controlled.

NOTE Vapor pressure is the *force* that comes off a material.

Several other important factors related to vapor pressure and volatility, such as temperature, surface area, wind, and surface type, require discussion. The figures usually provided for vapor pressure or volatility are standard at 68°F, unless indicated otherwise. If local conditions differ from that temperature, then the material will react in relation to the temperature: The cooler the temperature, the less its vapors are going to be an issue, and the hotter the temperature, the more vapors will be produced. Surface area plays a factor in the rate of evaporation, as shown in Figure 1-8, as the smaller the surface area, the lesser the amount of vapors that will be produced. Air movement or wind conditions also play a factor because greater air movement results in a quicker move to the gaseous state. One side note to consider with air movement is the vapor density of the product and its ability to be carried up and away. The last factor is the type of surface the release is on. There is a difference between concrete and sand, and chemicals evaporate differently from different surfaces.

NOTE Volatility is the relative *quantity* of the vapors that come from the material.

A material is considered an inhalation hazard if it has a vapor pressure in excess of 40 mm Hg, at which point the amount of risk a chemical presents increases. Chemicals with a vapor pressure in excess of 40 mm Hg have greater potential to move away from the immediate site of the event, can be flammable, or require increased isolation and/or evacuation distances. A chemical that has a vapor pressure of less than 40 mm Hg, although perhaps extremely toxic or corrosive, usually only harms through touch or ingestion.

SAFETY A material is considered an inhalation hazard if it has a vapor pressure in excess of 40 mm Hg.

TABLE 1-3

Chemical and Physical Properties of Selected Chemicals

Chemical	Molecular Weight	Freezing (°F)	Boiling Point (°F)	Flash Point (°F)	Vapor Pressure (mm Hg)	Autoignition Temp (in air)	LEL (%)	UEL (%)	IP Ionization Potential (eV)
Acetone	58.1	–140	133	0	180	869	2.5	12.8	9.69
Acetylene	26	–119	Sub.	Gas	33440	581	2.5	100	11.4
Ammonia	17	–108	–28	Gas	6460	1204	15	28	10.18
Arsine	78	–179	–81	Gas	11400		5.1	78	9.89
1, 1 dimethylhydrazine	60.1	–72	147	5	103	480	2	95	8.05
Dimethyl sulfate	126.1	–25	370	182	0.1	370			< 11.7
Ether	74.1	–177	94	–49	440	356	1.9	36	9.53
Ethion	384.5	10	302	349	0.0000015				
Fuel Oil # 2	~170	–238	347	126	5 (100°F)	494	0.7	5	< 10.6
Gasoline	~72	–267	102	–50	300	880	1.4	7.6	< 9.8
Hydrazine	32.1	36	236	99	10	74–518	2.9	98	8.93
MEK	72.1	–123	175	16	78	759	1.4	11.4	9.54
Toluene	92.1	–139	232	40	21	896	1.1	7.1	8.82
Triethylamine	101.2	–175	193	20	54	480	1.2	8	7.5
Xylene	106.2	–13	292	90	7	867	0.9	6.7	8.56

Figure 1-8 Vapor pressure and other factors play a role in how fast a chemical evaporates. (A) Wind conditions. (B) The temperature of the area around the spill and the temperature of the liquid. (C) Type of surface. (D) Surface area.

Vapor Density

Vapor density determines where vapors are located in relationship to air, as shown in Figure 1-9 and is discussed in the case study Hot Gasoline in this chapter. Air has a value of one. Gases with a vapor density of less than one will rise in air; gases with a vapor density of greater than one will sink in comparison to air. This property is important as responders need to know where to monitor to determine where the gases will be found. If the identity of the gas is unknown, then check low, then midway, and then upper levels of the area being sampled. Many reference texts provide vapor densities for gases, and for some liquids that move to the gaseous state. The **National Institute of Occupational Safety and Health (NIOSH)** *Pocket Guide* lists the vapor density as **relative gas density (RgasD)** for all of the chemicals occurring as gases. For other materials and for chemicals for which the vapor density is not provided, we can compare the **molecular weight** of the gas with the molecular weight of air. The molecular weight air is 29. Gases with a molec-

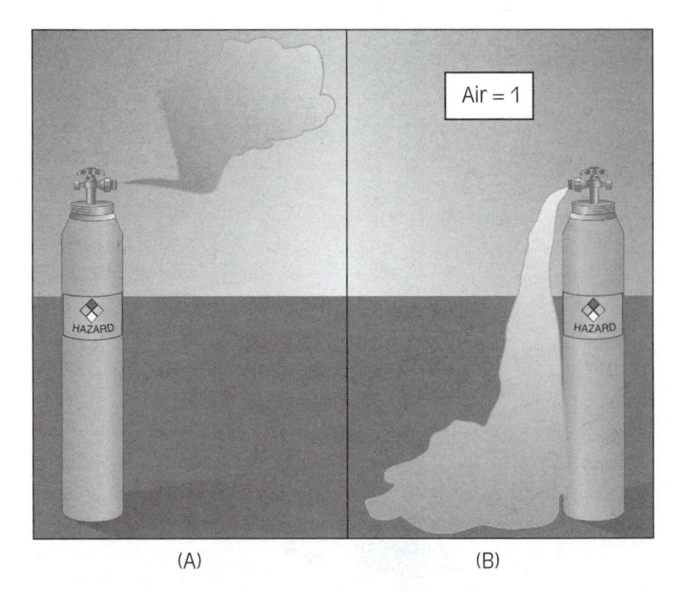

Figure 1-9 (A) Gases or vapors with a vapor density of less than 1 or with a molecular weight of less than 29 will rise in comparison to air. (B) Gases or vapors with a vapor density of greater than 1 or a molecular weight of more than 29 will sink.

TABLE 1-4

Common Gases That Rise

Gas Name	Vapor Density	Gas Name	Vapor Density
Diborane	0.97	Ammonia	0.60
Methane	0.55	Neon	0.7
Hydrogen	0.1	Helium	0.138
Hydrogen cyanide	0.9	Carbon monoxide	0.97
Acetylene	0.91	Ethylene	0.98

ular weight of less than 29 will rise and greater than 29 will sink in comparison to air. For street purposes, when dealing with an unidentified gas, most are low in comparison to air. Only ten common gases rise in comparison to air. The mnemonic that is used to remember them is DAMN 3H CAE, or DAMN 3H Cows Ate Everything. A common mnemonic for lighter-than-air gases, used to be HAHA MICEN, but this left out hydrogen cyanide and diborane gases. The ten gases for DAMN 3H CAE are listed in Table 1-4. Chemicals that have vapor densities close to the factor of 1 (including 1) such as 1.1, 1.2, 0.9, and 0.8 hang at midlevel and do not go anywhere, unless there is wind to move them along.

Weather affects vapor density and has a dramatic impact on how a gas reacts. The vapor densities provided in Table 1-4 are at standard temperature and pressure, as would be considered an average day. Natural gas has a vapor density of 0.6, which means that it should go up. On days when the humidity is high or there is ground fog, natural gas has a tendency to stay low. Propane, which has a vapor density of 1.6, is a real risk because it likes to stay low, but on hot dry days with lots of sun it will rise. These weather conditions are the extreme in most areas of the country, and when dealing outside the norm, do not expect chemicals to react in their normal fashion. This is another reminder that science is not black and white, there are considerable shades of gray.

NOTE Only ten common gases rise in comparison to air.

Accuracy and Precision

Two terms that are two different characteristics are commonly used to describe air monitors: **accuracy** and **precision.** Accuracy is the ability of the meter to

CASE STUDY

Hot Gasoline—A gasoline tank truck struck the underside of a beltway bridge, killing the driver and releasing the contents of the two front compartments. First-arriving crews did a great job of protecting the bridge and setting up foam lines to extinguish the tanker. Then the gasoline had to be transferred from the three remaining compartments. This was an interesting undertaking as the gasoline was still very hot from the intense fire. During the transfer, the transfer hoses were hot to touch even with gloves on. Air monitoring was essential to track the vapors that were being expelled by the transfer operation. True to form, the vapors stayed low and followed the natural lay of the land. Large isolation areas were set up to keep personnel out of the danger areas. The use of foam is always tricky as there has to be a balance between vapor suppression and making the environment safe to work in. The more foam there is and the more places it is applied, the more the work areas are slippery and difficult to work in. Air monitors can assist in determining when foam needs to be reapplied. If you start to get LEL readings when monitoring, that means the foam layer is breaking down releasing vapors, and foam needs to be reapplied. Some texts state that foam ingredients have a tendency to cause damage to the LEL sensor, but this only occurs over a long-term exposure, and not in the few hours that they would be used at a tanker incident. The best thing to do after an incident of this type is to calibrate the monitor, so as to ensure precision and accuracy.

produce findings as close as possible to the actual quantity of gas. A meter that is considered accurate when it is exposed to 80 ppm of a known gas should read 80 ppm. A meter that is not accurate may read 30 ppm instead of the 80 ppm exposure. Precision describes the ability of the monitor to reproduce the same results each time it samples the same atmosphere. In other words, it duplicates the readings each time the same concentration of gas is sampled. If a monitor is exposed to 70 ppm of carbon monoxide, one would expect the monitor to display 70 ppm, or at least close to that. If the monitor displays 40 ppm with every sample, it would still be precise, just not accurate. A meter that reads 64, 81, and 60 for the same 70 ppm sample is more accurate, as it is providing values close to the actual 70 ppm sample, but is not precise due to the variety of the readings. Precision refers to the repeatability of the readings. An air monitor can be precise, but inaccurate and vice versa. See Table 1-5 for more examples.

Many factors affect precision and accuracy, including, but not limited to, chemical and physical properties of the sample, weather (humidity and temperature are the two biggest factors), and the sensor technology.

Bump Tests and Calibration

Two other terms that need explanation are **bump test** and **calibration.** Also known as a field test, a bump test consists of exposing a monitor to known gases and allowing the monitor to go into alarm, thus verifying the monitor's response to the gas. It is not expected that the readings exactly match those of the bump gas, but they should be close. Bump gas comes in small disposable cylinders much like a spray can. For a four-gas mixture (LEL, O_2, CO, and H_2S), the cost is about $50 and has a shelf life of six months. One inexpensive method to bump test a monitor is to use a felt tip marker cap.

Calibration confirms that your monitor responds precisely to a known quantity of gas. New sensors usually read higher than they are intended; calibration electronically changes the sensor to read the intended value. As the sensor gets older, it becomes less sensitive and calibration electronically raises the value that the sensor displays. A sensor that cannot be electronically brought up to the correct value is dead and needs to be replaced. Calibrations used to be very difficult but with the new monitors and multigas cylinders it is relatively easy, depending on the manufacturer. The regularity of calibration is subject to debate, as some response teams calibrate daily, others every six months. The only item found in the regulations (for anything that requires air monitoring) is in the confined space regulation and that requires calibration according to the manufacturers' recommendations. Most of the written instruction guides from the manufacturers require calibration before each use. The definition of a calibration at this point is also subject to debate, as we can verify the monitor's accuracy by exposing it to a known quantity of gas, but not perform a "full" calibration. Most response teams establish a regular schedule of calibration (weekly or monthly) and then perform bump or field checks during an emergency response. Check with the manufacturer as to what calibration or bump test policy they recommend as they are all different. Calibration gas which has 4 gases (LEL, O_2, CO, H_2S) costs about $150.00 and will have a shelf life of about a year.

If you are concerned with potential liability, the best recommendation is to bump test one monitor, use it while performing a full calibration on the second unit, and then switch to verify the results of the first monitor. If you do not calibrate your instruments prior to an event, then immediately after the event calibrate them to ensure their accuracy and precision. Using a monitor that logs data internally assists in the record keeping necessary to protect

TABLE 1-5

Accuracy and Precision

Meter	Known Quantity of Gas	1st Reading	2nd Reading	3rd Reading	4th Reading	Meter Description
Meter A	100 ppm	50 ppm	50 ppm	50 ppm	50 ppm	Precise
Meter B	100 ppm	100 ppm	100 ppm	100 ppm	100 ppm	Accurate and precise
Meter C	100 ppm	70 ppm	90 ppm	85 ppm	95 ppm	More accurate than Meter A

yourself in court. Another method of providing protection is to keep good calibration records. If you compile regular calibration records, you can determine how much your monitor drifts from calibration to calibration. Once you have several months worth of data, you can be reasonably certain as to how accurate and precise your monitor actually is.

Lag and Recovery Time

All monitors have a certain reaction time or response time, sometimes known as **lag time.** Many manufacturers provide a response time, or the amount of time that the meter takes to report a reading that is 80 percent to 90 percent accurate. Some technical information separates lag time and response time, but both values represent a delay in the instrument providing a reading. The reaction time will vary depending on whether you are sampling with a pump. Monitors without a pump operating in a diffusion mode generally have a 15–30 second lag time. Monitors with a pump have a typical reaction time of 3–5 seconds, but even this can vary from manufacturer to manufacturer. Hand-aspirated pumps usually require ten to fifteen pumps to draw in an appropriate sample. The goal is to provide a given amount of volume of air across the sensors. When using sampling tubing, add 1 second of lag time for each foot of hose. Be sure to follow the manufacturer's recommended lengths of hose to ensure that the pump operates correctly. Only use hose provided by the manufacturer, as some hoses can absorb the chemicals you are sampling for and cause problems in future sampling efforts.

Monitors also have a **recovery time.** Recovery time is the amount of time that it takes the monitor to clear itself of the air sample. This time is affected by the chemical and physical properties of the sample, the amount of sampling hose, and the amount absorbed by the monitor.

Relative Response

Monitors are calibrated to a specific gas. Some LEL monitors are calibrated for methane (using pentane calibration gas) and are accurate and precise for only methane. Other monitors are calibrated using a pentane scale, using methane as the calibration gas. **Relative response** is a term used to describe the way the monitor reacts to a gas other than the one it was calibrated for. The monitor's manufacturer has tested the monitor against other gases and has provided a factor (relative response factor) that one can use to determine the amount of gas actually present when sampling. Some of these factors are provided in Table 1-6.

For example, suppose at a hexane spill you are using an ISC TMX-412 calibrated for pentane and the detector is reading 68% of the LEL. The response curve factor for hexane is 1.2, so you multiply this factor times the LEL reading.

Detector Reading × Response Curve Factor = Actual LEL Reading
$68 \times 1.2 = 82$ (rounded up)

So the actual LEL is 82 percent, a situation that is worse than what the instrument was providing. If you used the same instrument for a release of acetylene, the response curve is 0.7. Using the same scenario that the instrument was reading 68 percent of the LEL, the actual reading is 48% ($68 \times 0.7 = 48$), a safer situation than reported by the detector. It is not important to memorize response factors, but it is a good idea to have the response factors for the chemicals you commonly respond to listed on a laminated card attached to the monitor. The response curves that are important are those that are above one, which means that the atmosphere is worse than what your monitor is reading. If dealing with legal issues is a possibility, then it is important to do your math and calculate the response curves for the report. If the manufacturer does not provide response curves for the chemical you are dealing with, you can contact the manufacturer or you can make some estimations of the factor. For the most part, manufacturers' response curves do follow the molecular weight of the chemicals. It is not exact but can get you in the range of the factor. Looking at Table 1-6, we can see that the Industrial Scientific TMX-412 has response curve factors that track fairly close to the molecular weight, as do the MSA factors.

SAFETY The response curves that are important are those that are above one, which means that the atmosphere is worse than what your monitor is reading.

FORMULAS AND CONVERSIONS

A number of formulas and conversions apply in air monitoring. There is also a chart that provides some of the various chemical and physical properties of a given chemical. If you can identify at least one chemical and physical property, then you can use

TABLE 1-6

Response Curve Factors (calibrated to pentane)[a]

Gas Being Sampled	ISC TMX-412 Factors[b]	MSA 261 Factors[c]	RAE System Factors[d]	Molecular Weight
Hydrogen	0.5	0.6		1.0
Methane	0.5	0.6	1	16.04
Acetylene	0.7	0.8		26
Ethylene	0.7	0.8		28.06
Ethane	0.7	0.7		30.08
Methanol	0.6	0.7		32.1
Propane	0.8	0.9	1.88	44.1
Ethanol	0.8	N/A	1.69	46.1
Acetone	0.9	1		58.1
Butane	0.9	1		58.1
Isopropanol	1	1.1		60.1
Pentane	1	1	1	72.2
Benzene	1	1.1	2.51	78.1
Hexane	1.2	1.3		86.2
Toluene	1.1	1.2	2.47	92.1
Styrene	1.1	N/A		104.2
Xylene	1.3	N/A		106

[a]Always use the response factors supplied by the manufacturer, keeping in mind that these are laboratory estimations.
[b]Factors for Industrial Scientific Corporation LEL sensor 1704A1856-200 calibrated with pentane.
[c]Factors for the Mine Safety Appliances MSA 360/361 calibrated with pentane.
[d]Factors are for the RAE system catalytic bead LEL sensor calibrated with pentane.

this chart to compare it to other known chemicals. Gasoline and diesel are two chemicals that provide a good comparison for this chart. Table 1-7 provides the interrelationships of some chemical properties. Table 1-8 provides some commonly used conversions and formulas.

TABLE 1-7

Chemical Property Interrelationships

If the chemical has:	It then has a low:	It also has a high:	It has a wide:	It has a narrow:
Low molecular weight	Boiling point Flash point Heat output	Vapor Pressure Evaporization rate Ignition temperature	Flammable range	
High molecular weight	Vapor Pressure Evaporization rate Ignition temperature	Boiling point Flash point Heat output		Flammable range

TABLE 1-8

Common Conversions and Formulas	
Conversion or Formula Name	**Conversion or Formula**
Celsius to Fahrenheit	(°C × 1.8) + 32 = °F
Fahrenheit to Celsius	(°F – 32) ÷ 1.8 = °C
Percent volume to ppm	1 percent volume = 10,000 ppm
ppm to mg/m^3	$\text{Mg/m}^3 = \dfrac{\text{ppm} \times \text{molecular weight}}{24.5}$
Mg/m^3 to ppm	$\text{ppm} = \dfrac{(\text{mg/m}^3)\,(24.5)}{\text{molecular weight}}$
Vapor density	Molecular weight ÷ 29 = Vapor Density
Vapor density in high humidity	Molecular weight + 18 ÷ 29 = Vapor Density
Volatility	$V = \dfrac{16020 \times \text{Molecular Weight} \times \text{Vapor Pressure}}{°\text{Kelvin}}$
°Kelvin	0°C = 273 Kelvin, so 68°F (20°C) is 293 Kelvin

SUMMARY

The basic terminology and chemical properties are key to understanding air monitoring. Each term discussed in this chapter is used in later chapters, so it is essential that you understand them. With the assistance of air monitoring and with some background knowledge of chemical properties, some assessments can be made as to the type and severity of the event. All air monitors are essentially dumb devices that take readings and output a number. It is up to the human using the device to interpret that number and make an educated assumption as to what it means. The chemical and physical properties are important in making that assumption, and to the eventual mitigation of the event.

KEY TERMS

Accuracy Used to describe a monitor that is able to provide readings close to the actual amount of gas present.

Boiling point Temperature at which a liquid changes to a gas. The closer to the boiling point, the more vapors that are produced.

Bump test Using a quantity and type of gas to ensure that a monitor responds and alarms to the gases being tested for.

Calibration Checking the response of a monitor against known quantities of a sample gas, and if the readings differ, then electronically adjusting the monitor to read the same as the test gas.

Chemical Abstract Service A service that registers chemical substances and issues them a unique registration number, much like a social security number. Also known as CAS.

Fire point The lowest temperature at which a liquid ignites and with an outside ignition source sustains burning.

Flammable range The numeric range, between the lower explosive limit and the upper explosive limit, in which a vapor will burn.

Flash point The minimum temperature of a liquid that produces sufficient vapors to form an ignitable mixture with air when an ignition source is present above the liquid.

Hazardous Waste Operations and Emergency Response An OSHA regulation that covers waste site operations and emergency response to chemical emergencies.

Lag time The amount of time it takes for a monitor to respond once exposed to a gas.

LEL See Lower explosive limit.

Lower explosive limit The least amount of flammable gas and air mixture in which there can be a fire or explosion.

Molecular weight Weight of the molecule based on the periodic table, or the weight of a compound when the atomic weights of the various components are combined.

National Fire Protection Association A consensus group that issues standards related to fire, HAZMAT, and other life safety concerns.

National Institute of Occupational Safety and Health The research agency of OSHA, which studies worker safety and health issues.

NFPA See National Fire Protection Association.

NIOSH See National Institute of Occupational Safety and Health.

Occupational Safety and Health Administration Government agency tasked with providing safety regulations for workers.

OSHA See Occupational Safety and Health Administration.

Precision The ability of a detector to repeat the results for a known atmosphere.

Recovery time The amount of time it takes for a detector to return to zero after exposure to a gas.

Relative gas density Term used by the NIOSH *Pocket Guide* that means vapor density. It is a comparison of the weight of a gas to the weight of air.

Relative response How a monitor reacts to a given gas as compared to the gas for which the monitor was calibrated.

Risk-based response A system for identifying the risk chemicals present even though their specific identity is unknown. This system characterizes all chemicals into fire, corrosive, or toxic risks.

UEL See Upper explosive limit.

Upper explosive limit The upper range of the flammable range; the maximum concentration of flammable gas or vapors that can be present mixed with air to have a fire or explosion.

Vapor density (VD) The weight of a gas compared with an equal amount of air. Air is given a value of 1, and gases with a vapor density less than 1 will rise, while those with a VD of greater than 1 will sink.

Vapor pressure The force of vapors coming from a liquid at a given temperature.

Volatility The amount of vapors coming from a liquid.

REFERENCES

National Institute for Occupational Safety and Health and the Centers for Disease Control. 1999. *NIOSH Pocket Guide to Chemical Hazards*. Washington, DC.

NFPA 471. 1997. *Recommended Practice for Responding to Hazardous Materials Incidents*. Quincy, MA.: National Fire Protection Association.

NFPA 472. 1997. *Professional Competence of Responders to Hazardous Materials Incidents*. Quincy, MA.: National Fire Protection Association.

IDENTIFYING THE CORROSIVE RISK

 INTRODUCTION

Because humans and electronic devices usually are harmed when exposed to corrosive materials, determining the pH of a released material is very important. The reasons are obvious: Corrosive materials will burn or irritate humans and render any electronic instrument useless fairly quickly. Most electronic detection devices will work for a limited time in a corrosive atmosphere, but eventually stop working, sometimes unknown to the user. **pH paper** is one of the most useful tools that a responder can depend on to identify a corrosive risk. It is recommended that pH paper be one of the first items available at a chemical release and the easiest way to do that is to attach a strip of pH paper to the four-gas instrument. Also, determining the pH of a **hydrolysis material** is one of the methods of detecting warfare agents, as described in the case study Explosives in the Basement.

SAFETY Because humans and electronic devices usually are harmed when exposed to corrosive materials, determining the pH of a released material is very important.

Fortunately for emergency response, determining and interpreting pH is fairly easy. The most common method is to use a multirange pH paper, shown in Figure 2-1, which is available in various

Figure 2-1 Multirange pH paper measures the corrosive risk.

CASE STUDY

Explosives in the Basement—The police department was called by a car wash that was under explosive attack. When the police officer arrived at the car wash, he too was taking explosive rounds bursting in the air. He determined the trajectory of the devices and went to that location. When he went into the backyard he noticed a fiftyish man with a smoking polyvinyl chloride (PVC) pipe in his hand. When asked by the officer where he got the fireworks, the obviously intoxicated man—not recognizing that the person he was talking to was a police officer—replied "Come on, I will show you." When they went into the basement of his house, there were explosives, bomb-making supplies, and many ash trays and empty beer cans all over the place. The police officer arrested the gentleman and retreated out of the basement, calling for the bomb squad. When it arrived, the bomb squad needed help identifying some of the chemicals in the basement. Although there were many things to be concerned with throughout the basement, some very serious explosives included pentaerythritol tetranitrate (PETN) in the freezer. There were a number of corrosives, including concentrated nitric acid. We tested the pH and it was about 0, and then using a Spilfyter™ chemical classifier (described in Chapter 9) identified that it was also an oxidizer, exhibiting the characteristics of nitric. The bomb technicians gathered up the explosives, and we gathered up the many chemicals for disposal.

sizes and ranges. It has a range of 0–13 pH and detects both liquids and gas corrosives. Wetting the paper with neutral water makes it more sensitive to gaseous corrosives. For emergency response this multirange paper is more than sufficient, as it is not necessary for us to determine pH down to the tenths; whole numbers are more than adequate for a quick assessment.

NOTE Most electronic detection devices will work for a limited time in a corrosive atmosphere, but eventually stop working.

pH is based on the percent of hydrogen ion concentration using a logarithmic scale, with acids being between 0 and 7. Bases are between 7 and 14 with 7 being neutral. In legal terms (by no means chemically), neutral in most states is between 5 and 9 and such materials can be safely disposed of without any permits or special conditions. When we measure the pH, we are actually measuring the number of free hydrogen or hydroxyl ions in the corrosive, which is sometimes referred to as *strength.*

Another factor that must be considered when dealing with corrosives is **concentration.** Most corrosives have some amount of water in them, and some are diluted to low percentages. As an example, sulfuric acid comes in a range of concentrations from 2 percent (spent pickle liquor), to the normal range of 30 to 37 percent for battery electrolyte to 98 percent lab and commercial concentration. One unusual form of sulfuric acid is oleum, which is concentrated sulfuric acid that exists as 120 to 160 percent sulfuric acid. Oleum is supersaturated with sulfur trioxide. Normally sulfuric acid does not have much vapor pressure (1 mm Hg at 294°F, or 0.001 mm Hg at 68°F), but oleum, also called *fuming sulfuric acid,* has a high vapor pressure. When it is released, it interacts with water vapor in the air and creates a significant vapor cloud of concentrated sulfuric acid as shown in Figure 2-2.

SAFETY Chemicals with a pH of less than 2 or more than 12 present a significant risk for injury to humans.

Hydrochloric acid (HCL) is commonly found at 20 to 25 percent and is known as muriatic acid (brick cleaner). It is available commercially at 32 to 37 percent. Removing relative concentration from the equation and using only pH strength, we can look at the corrosives that have potential for immediate harm. Chemicals with a pH of less than 2 or more than 12 present a significant risk for injury to humans. Not all contact with material with a pH of 0 causes immediate skin damage. In many cases the skin will only be irritated. In some cases though, as with oleum, sodium hydroxide, or potassium hydroxide, the contact results in a burn within minutes and requires quick washing to prevent further damage. Having hydrochloric or sulfuric splashed on you results in skin irritation and burns after a few minutes, but quick washing will minimize or prevent the damage as related in the case study Garbage Truck Versus Van. Time is another factor in dealing with corrosives, as after a period of time a material

Figure 2-2 Training involving oleum (concentrated sulfuric acid), a very aggressive corrosive with high vapor pressure. (Photo courtesy of Buzz Melton.)

with a pH below 2 or above 11 will cause damage to skin. The expression of time in this instance, however, is in hours. However, any corrosives in the eyes can cause immediate damage and possibly a permanent loss of vision. Examples of a common corrosive are Pepsi™ and Coke™ which have a pH of 2.5, which we can drink, but over a period of time can cause damage to skin or painted surfaces such as on a car.

SAFETY Any corrosives in the eyes can cause immediate damage and possibly a permanent loss of vision.

CASE STUDY

Garbage Truck Versus Van—A HAZMAT assignment was dispatched to an auto accident with entrapment, in which a Dumpster-style garbage truck had been involved in a head-on collision with a van. The van was marked as a carpet-cleaning van, and had two occupants who were well entangled in the wreckage. The forks had been down on the garbage truck, complicating the rescue. The rescue crews who were working to free the victims said the passenger was complaining of his face burning. The rescue crews, all in full PPE, looked in the back of the van where all the cleaning chemicals had been thrown around, many of them leaking. We figured that some of the cleaning chemicals had been splashed on the passenger. We checked the pH of the victim, which was severely acidic, while the pH of the mixture in the back of the van was on the opposite end of the scale to the caustic side. We quickly removed the victim and began the decon process, taking care of his head and face. We continued to check the products in the back of the van, looking for the acid so we could let the hospital know what material had been spilled on the patient. Unable to locate any acids, we started looking around the front of the van and in the garbage truck. We determined that one of the forks from the garbage truck had gone through the battery on the passenger side of the van, splashing battery acid on that side of the van.

METHODS OF pH DETECTION

pH paper, often incorrectly referred to as litmus paper, comes in several different forms, but the most common is 1/4-inch rolls that have a pH range of 1–12 or 0–13. Also available are individual strips that cover the range of 0–14, as well as individual numbers. The multirange paper is the easiest and quickest to interpret. Individual strips with a multitude of colors to match is more difficult to correctly interpret. It is advisable to test your pH paper for its effectiveness, as it varies from brand to brand among the color ranges. For example the pHydrion™ 0–13 range is much quicker to respond to gases than the 1–12 range paper.

The most common paper has a range of 0–13 and has one-step increments based on a color change. Paper is available in smaller ranges (1–3 for example) and with increments such as 1.1, 1.2, 1.3 and so on. Multicolor strips are made that offer the same ranges as the paper rolls and offer some additional reliability. It is not generally important to have pH paper that tells the pH is 1.3 when a less expensive roll let us know that it has a pH of 1. For emergency response it is not necessary to have the exact pH of a spilled material, but responders do need to know if it is corrosive or neutral. It is also necessary to know whether the material has a corrosive vapor or is just a liquid corrosive.

Another method of determining pH of materials is to use a pH probe, which is commonly used in labs. These probes start at one hundred dollars and run into several thousand dollars. They are not commonly used in emergency response as the sensors need to be replaced regularly (6 months to a year), are costly, and are easily broken. The sensors usually have to be kept wet and are ruined when dried out. The sensors must be calibrated before each use, so they must be maintained with the appropriate calibration solutions. The sensors are shock-sensitive and are easily broken during the rigors of field use. The only time this type of device might be useful is when sampling a large number of materials in a fairly controlled environment. However, the pH probe requires cleaning after each sampling prior to measuring another sample.

Because pH paper turns colors in the presence of corrosive vapors, it can be used to indicate corrosive atmospheres as well as liquids. Although dry paper is a pretty good indicator of the presence of corrosive gases, wetting the paper makes it react more quickly, especially when lower amounts of corrosives are in the air. Make sure that neutral water is used to wet the paper. Although some sources recommend that you use deionized water, for street purposes, neutral water suffices. Using a setup as shown in Figure 2-3 provides protection against corrosive gases and does not require handling. The X or T format allows for three or four liquid samples to be taken with one pair of forceps, but if vapors are present, all of the paper may change color.

In some cases the liquids you are sampling will present misleading readings on pH paper and may be misinterpreted. When reading pH paper, you not only need to identify the color change but also interpret the leading edge of the paper. This is usually only a concern when sampling liquids you think have a pH of 4 to 8. All of the other pHs are extreme enough that their results usually are not misinter-

CASE STUDY

Contagious Asthma Patients—While headed to lunch one day, my boss and I heard a dispatch for two kids having an asthma attack. Being inquisitive and wondering why two kids would be having simultaneous asthma attacks at an elementary school, he called dispatch for more information. Dispatch did not have much additional information, other than the kids were in a classroom and started to have an asthma attack, and now a third was also having trouble. Since the cause was probably chemical-related, a HAZMAT assignment was requested. When crews arrived, they found the kids had been evacuated into the school yard. The teacher of the classroom in question was asked about possible causes. After much questioning it was determined that the students started having problems after the teacher sprayed a cleaning solution on the students' tables. She did not know what kind it was, but could show us the bottle, which was a household cleaning solution. We asked if the label matched the contents, and she said that the janitor would fill it as needed. The janitor said he filled the bottle from a 55-gallon drum in the shop. He did not have a material safety data sheet (MSDS) sheet for the solution, but the drum was labeled. We asked if he diluted the cleaner in the spray bottle, to which the answer was no. In the drum was pure potassium hydroxide, a very serious corrosive with a pH of 13, which obviously and understandably irritated the students.

Figure 2-3 By using a T or an X formation, you can accomplish three or four tests with one pair of forceps.

preted. The problem is usually with hydrocarbons, which are neutral substances, but when sampling for pH may cause an erroneous reading of the pH paper. When multirange pH paper is wetted with a substance such as a hydrocarbon, it gives the impression that it has a pH of 4 to 8. The paper is just wet and really did not change. Two methods can be used to determine if the pH is accurate. If the leading edge for a material that is indicating a pH of 4 to 8 and is straight, as shown in Figure 2-4, the material is most likely neutral. This test is not 100 percent but offers fair reliability. The other test, which is 100 percent accurate, only works with high vapor pressure materials. When sampling, stick the pH paper into the vapor space of the sample jar, and check for a change. If there is a change in the vapor space, then check the liquid's pH and record those results as this material is a corrosive. If the pH paper did not change, but in the liquid you are interpreting the readings have a pH of 4 to 8, then record those readings and complete an evaporation check. With a pipette, take a small sample from the sample jar. Place one drop on a watch glass, start a clock, and watch the sample. If the sample evaporates in 5

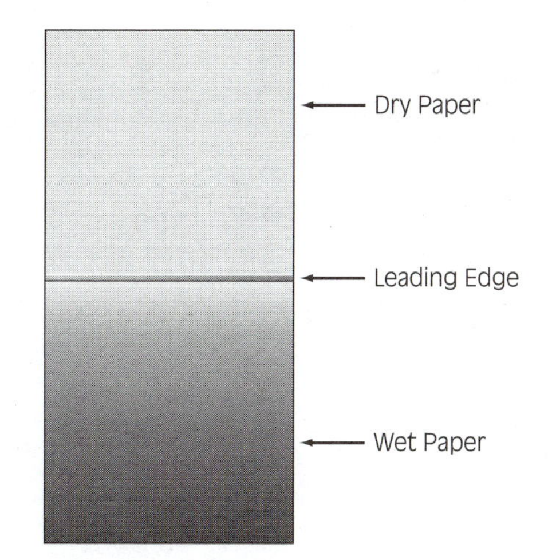

Figure 2-4 When the pH is reported to be between 4 and 8 and has a straight leading edge, this liquid is most likely a neutral substance.

minutes or less, then the material is neutral. A material that evaporates in 5 minutes has a high vapor pressure, and so would have changed the pH paper in the vapor space. If it did not, the pH reading was from wet pH paper. A leading edge on pH paper that is jagged, multicolored, and seems to be wicking through the pH paper is a corrosive and should be recorded as such. This leading edge is shown in Figure 2-5. The pH of common materials is provided in Table 2-1.

NOTE When reading pH paper, you not only need to identify the color change but also interpret the leading edge of the paper.

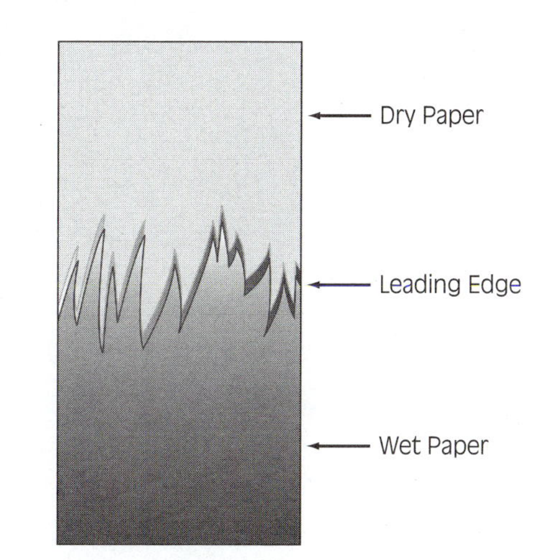

Figure 2-5 When the pH is reported to be between 4 and 8 and the leading edge is jagged, this substance is a corrosive at the pH read on the strip.

TABLE 2-1

pH of Common Materials

Material	pH	Material	pH
Orange juice	3	Sulfuric acid	0
Ammonia	13	Pepsi™	2–3
Vinegar (acetic acid)	3	Hydrochloric acid	0
Sodium hydroxide (lye)	13	Oleum	1
Acetone	7	Gasoline	7
Hydrofluoric acid	0–1	Diesel fuel	7

SUMMARY

Testing for corrosives is a very important activity, not only for responder safety, but for the preservation of the instruments. Corrosives are the second most released substance behind flammables and combustibles, so you must be comfortable with pH sampling. Practice with known materials and you will become familiar with how common corrosives react.

KEY TERMS

Concentration The amount of corrosive as compared to water in a corrosive substance.

Hydrolysis material The breakdown product(s) of a material.

pH paper Testing paper used to indicate the corrosiveness of a liquid.

OXYGEN LEVEL DETERMINATION

- Introduction
- Monitoring for Oxygen
- Oxygen Monitor Limitations
- Summary
- Key Terms

INTRODUCTION

SAFETY In oxygen extremes (deficient/enriched) there is a major problem as some gas (which may include oxygen) is present in large quantities.

After pH, oxygen is the next most important thing to sample for, as humans need it to survive and the monitoring instruments need it to function correctly. Normal air contains 20.9 percent oxygen; below 19.5 percent, or **oxygen deficient,** is considered to be a health risk, and above 23.5 percent, or **oxygen enriched,** is considered a fire risk. If an oxygen drop is noted on the monitor, one or possibly more than one contaminant is present, causing the reduced oxygen levels and another hazard (i.e., toxic, flammable, corrosive, or **inert**) is causing the oxygen-deficient atmosphere. In oxygen-deficient atmospheres, any combustible gas readings are also deficient and cannot be relied on. In an oxygen-enriched atmosphere, the combustible gas readings increase and are not accurate.

NOTE Every 0.1 percent drop in oxygen on the meter was worth 5,000 ppm of something else, which is a considerable amount of a potential toxic material in the air.

SAFETY When oxygen sensors are turned on, the readings will fluctuate and in some cases provide erroneous readings until the electronics are warmed up.

MONITORING FOR OXYGEN

In oxygen extremes (deficient/enriched) there is a major problem because a considerable amount of some gas (may include oxygen) is present. When there is a drop in oxygen, the amount of occupying gas can be calculated with some known information. For the sake of this scenario the calculations are rounded off. In air we have 79 percent nitrogen (N_2), which equates to 790,000 ppm, and roughly 21 percent oxygen (O_2), which equates to 210,000 ppm oxygen. Using a rough rule of fifths, the makeup of air is four-fifths N_2 and one-fifth O_2. If it is assumed there is a contaminant that will drop the O_2 content by 5,000 ppm, it can then be assumed that the N_2 will drop by 20,000 ppm (four times as much stuff in air to drop N_2), so the total amount of stuff in the air is 25,000 ppm or 2.5 percent by volume. The resulting drop in the oxygen level would be to 20.5 percent (using 21 percent as our starting point), which means that every 0.1 percent drop of oxygen on the meter was worth 5,000 ppm of something

Figure 3-1 Typical oxygen sensors.

else—a considerable amount of a potential toxic material in the air. So a drop in oxygen that goes from 20.9 percent to 20.7 percent is significant, and responders should wear SCBA as a minimum. One warning, though, is that when oxygen sensors are turned on, the readings will fluctuate and in some cases provide erroneous readings until the electronics are warmed up. The meter may indicate its readiness, but until the unit is warmed up and running for seven to ten minutes, the reading will fluctuate, particularly in cold winter conditions.

OXYGEN MONITOR LIMITATIONS

Most oxygen sensors only last a year or two, because they are always working by being exposed to oxygen in the air. Many oxygen sensors are now warranted for two years, and unless exposed to high concentrations should last the two years. An oxygen sensor (Figure 3-1) is an electrochemical sensor that has two electrodes within a gellike material as shown in Figure 3-2. When oxygen passes through the sensor, it causes a chemical reaction, creating an electrical charge and causing a readout on the monitor. As long as the sensor is exposed to oxygen, it will cause a reaction within the electrolyte solution sealed within the sensor. Even with the instrument turned off, the sensor is being exposed to oxygen and is working. This sensor is the most often replaced, usually every one to two years. Exposures that hurt oxygen sensors are chemicals with lots of oxygen in their molecular structure, including carbon dioxide (CO_2) and strong oxidizing materials, such as chlorine and ozone. The problem with CO_2 is that it is always present in the air, and the higher the percentage, the faster the sensor will deteriorate. High humidity and high or low altitude also affect oxygen sensors.

NOTE Exposures that hurt oxygen sensors are chemicals with lots of oxygen in their molecular structure, including carbon dioxide (CO_2) and strong oxidizing materials, such as chlorine and ozone.

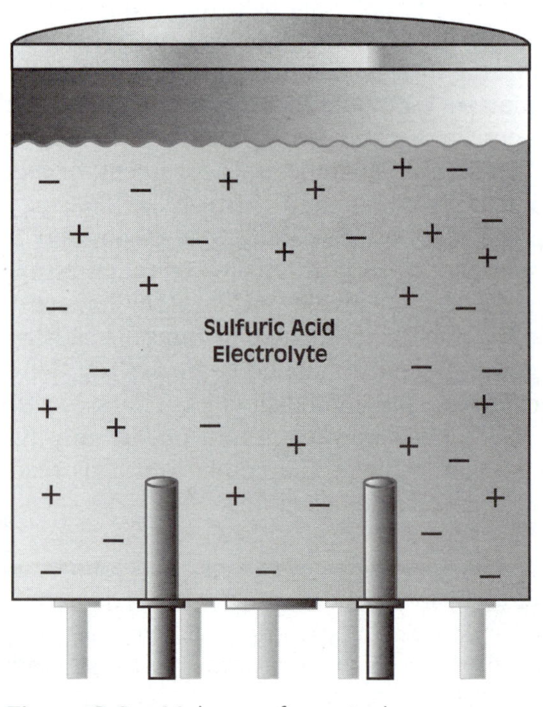

Figure 3-2 Makeup of a typical oxygen sensor.

The optimal temperature for operation is between 32°F and 120°F. Between 0°F and 32°F the sensor slows down. Temperatures below 0°F can permanently damage the sensor, usually freezing the sensor causing it to rupture. The operation depends on absolute atmospheric pressure and calibration to the atmospheric pressure responders will be sampling. Responders should calibrate the sensor especially at the temperature and pressure (altitude) of the area they will be sampling in.

SUMMARY

Measuring oxygen is of prime importance, not only because we need oxygen to survive, but because most monitors also require oxygen to function. The fact that an atmosphere may be oxygen deficient provides significant clues as to the potential for large amounts of a toxic or asphyxiating substance being present. Being in an oxygen-enriched atmosphere presents a greater fire risk and usually means that free oxygen or an oxidizer is involved in a chemical reaction.

KEY TERMS

Inert A chemical that is not toxic, but that displaces oxygen.

Oxygen deficient An oxygen level below 19.5 percent.

Oxygen enriched An oxygen level above 23.5 percent.

FLAMMABLE GAS DETECTION

- Introduction
- Combustible Gas (LEL) Sensor Types
- Linearity
- Summary
- Key Terms

INTRODUCTION

Combustible gas indicators (CGIs), also referred to as combustible gas sensors, have been used by the fire service and industry for many years. The term *combustible gas indicators* does not accurately describe what they detect. In reality they only measure flammable gases, not combustible gases. A more accurate name for these meters is **LEL meter** and the actual sensor that does the detecting should be called an **LEL sensor,** terms that are used throughout the remainder of this text. Most of the new LEL meters are used to measure the lower explosive limit (LEL) of the gas for which they are calibrated. Figure 4-1 shows a common meter. LEL meters are used to identify the fire risk in the risk-based response system. Most LEL meters are calibrated for methane (natural gas) using pentane gas. When calibrated for methane, the LEL sensor will read up to the LEL; some new units will shut off the sensor when the atmosphere exceeds the LEL. This is an important consideration, because the longer the sensor is exposed to an atmosphere above the LEL, the faster it will deteriorate. LEL sensors are available that read above the LEL by reading in percent by volume. These sensors are known as volume or concentration devices. Landfills and some industrial applications need LEL monitors that read 100 percent by volume because many times these applications have methane in quantities that greatly exceed the LEL, and they need to monitor the high levels.

Figure 4-1 A typical LEL monitor that is part of a four-gas detection device.

Although having this capability has advantages, it really does not matter whether the atmosphere is 100 percent of the LEL or 70 percent by volume. Once the LEL is exceeded, there is a fire hazard. Responders' tactics will not change because the volume LEL meter read 74.5 percent by volume and the LEL monitor read 100 percent of the LEL.

NOTE The term *combustible gas indicators* does not accurately describe what they detect.

The LEL meter reads up to 100 percent of the LEL, which for methane is 5 percent. Thus, if an LEL meter calibrated for methane reads 100 percent, the actual concentration by volume of methane is 5 percent. If the LEL meter reads 50 percent, then the concentration for methane is 2.5 percent.

SAFETY LEL meters are used to identify the fire risk in the risk-based response system.

Any flammable gas sample that passes over the sensor causes a reaction. How much of a reaction depends on the gas. Each LEL meter comes with a relative response curve for other gases. The exact number of other gases referenced varies from manufacturer to manufacturer. Response curves are discussed in Chapter 1.

Both OSHA and the Environmental Protection Agency (EPA) have established action levels shown in Table 4-1 that provide a safe level expressly because of the relative response curve problem. The OSHA values are from the confined space regulation (29 CFR 1910.146) and are only applicable in confined spaces, although many response teams have adopted those values for standard HAZMAT work. They have factored the various response curves in and provide guidelines that can be followed.

SAFETY If you obtain a reading of 1 percent on your properly calibrated LEL meter at the door to a building, you may have a problem.

The unfortunate thing about the EPA action guidelines is that they were originally designed for hazardous waste sites (which are usually outside) and they do not apply to a lot of what the fire service does, as in one case described in the Nitrocellulose case study. Some factors need to be established that would impact these action guidelines, because in some cases a 26 percent reading may require action. If you respond to a building for a natural gas leak and the meter reads 26 percent, this situation should be considered dangerous. The determining factor is where this reading was obtained. If it was obtained at the front door and you have not identified the source, then you may be in serious trouble. If the highest reading you obtain is 26 percent and you have located the source of the leak (and hopefully shut the gas off and have ventilated the building), you are in a relatively safe situation. To actually have a fire, the meter needs to read 100 percent, so in each instance you were 74 percent in the good range. If there is no other life hazard, then at 26 per-

TABLE 4-1

OSHA and EPA Air Monitoring Action Levels

Atmosphere	Level	Agency	Action
Combustible Gas	<10% LEL	EPA	Continue to monitor with caution.
	10–25% LEL	EPA	Continue to monitor, but use extreme caution especially as higher levels are found.
	>10%	OSHA	Requires evacuation of the confined space.
	>25%	EPA	Explosion hazard, evacuate the area.
Oxygen	<19.5%	EPA	Monitor with SCBA; CGI values are not valid.
	19.5%–25%	EPA	Continue monitoring with caution; SCBA not needed based on O_2 content only
	19.5%–23.5%	OSHA	Continue to monitor.
	>25%	EPA	Explosion hazard, withdraw immediately.
	>23.5	OSHA	Explosion hazard, withdraw immediately.

CASE STUDY

Nitrocellulose—A HAZMAT assignment was dispatched to a industrial facility. In the previous year the HAZMAT team had responded to the facility for an underground nitrocellulose tank that was leaking. Nitrocellulose is also known as gun cotton, the main component of Sterno™. That event was not particularly significant, as it involved the nitrocellulose bubbling up out of the ground into an adjacent parking lot. This call also involving nitrocellulose would be much more significant. Since the underground tank leak, the facility was getting nitrocellulose in 300-gallon **totes,** which were stored in an adjacent warehouse that also housed all of their hazardous waste. These drums of flammable waste were stacked from floor to ceiling, in addition to many other totes in the building. This building was adjacent to the main manufacturing facility and was just under the beltway overpass. When the HAZMAT crews arrived, they found that 300 gallons of nitrocellulose had been spilled out onto the warehouse floor, coating pallets and other drums and totes as seen in Figure 4-2. Upon arrival, the LEL readings were 90 percent of the LEL—a cause for concern. Because this liquid was flammable, foam was applied, which brought the LEL down to 85 percent. The HAZMAT team was concerned as it knew from the previous experience that nitrocellulose, usually stabilized with ether, alcohol, acetone and water, became a shock-sensitive explosive as it dried out. The team also knew that if it did ignite, it would be nearly impossible to extinguish without a large quantity of purple K extinguishing agent. There were some discussions about the cleanup and the contractor that would be needed to complete the job. The initial contractor balked at entering the atmosphere at 85 percent of the LEL, but the longer crews waited, the worse the explosive hazard would be, so cleanup was quickly initiated along with stabilization. It is always recommended to work in levels below 25 percent or 10 percent of the LEL, but this is not always possible.

cent you have no reason to be in the building and at 26 percent and you should evacuate. When dealing with unidentified situations, which includes natural gas leaks as they are unidentified until you confirm the source, you should consider a reading of 1 percent significant. If you obtain a reading of 1 percent on your properly calibrated LEL meter at the door to a building, you may have a problem, and there is potential that somewhere in that building the gas will be near the LEL. Keep in mind that 1 percent of the LEL equates to 10,000 ppm, which is a significant amount of a potentially toxic material. When dealing with life hazards, we really should not quantify safe or unsafe, because our obligation is to protect the public. Chapter 10 discusses risk and making life hazard and rescue decisions, but the

Figure 4-2 A nitrocellulose tote lost its valve, spilling 300 gallons on the warehouse floor. Quick action was required to minimize the extreme hazard presented by this material.

reality is that there cannot be a fire until the LEL meter reads 100 percent. Even when it reads 100 percent, people should be rescued as rapidly as possible. Anytime you move people they are in danger; when roads are shut down people are in danger for additional accidents. People panic and do not follow directions well. People will do everything in their power to escape a real or perceived threat. The best course of action in many cases is to remain sheltered in place. When there is no public life risk, then at 10 percent LEL responders should evacuate the building. When dealing with unidentified materials some additional considerations come into play, specifically response curve factors. The worst response curve for LEL sensors is a 5, which is rarely encountered, as all the others are below 2.6, which we will round up to 3 for ease of computation and to add a layer of safety. With a factor of 3 and an LEL meter reading of 33 percent, the actual atmosphere could be 99 percent. So when dealing with an unidentified gas, the action level should be 33 percent. By using other detection devices and identifying the chemical family, you may be able to increase this number to the lowest range for that family.

SAFETY When there is no public life risk, then at 10 percent LEL responders should evacuate the building.

COMBUSTIBLE GAS (LEL) SENSOR TYPES

The basic principle of most LEL sensors is that a stream of sampled air passes through the sensor housing, causing a heat increase, increasing the resistance in an electrical circuit, which when balanced by a known fixed resistance causes a reading on the instrument. The infrared sensor is the only one that does not follow that principle. There are four types of LEL gas sensors, and when purchasing or using a monitor it is important that responders know the sensor type. Readings can and do vary between the four, and responders' safety depends on that instrument so they must understand how it works. A common complaint at training is "When we used our meter next to the gas company rep's meter, his was going off and our was not reading at all." Or "His reading was twice what our meter stated." There are a couple of reasons for this. To compare readings you must compare sensor to sensor, and in many cases the gas company is using a **metal oxide sensor** (**MOS**) to look for small leaks, while most responders are using a catalytic bead sensor. There is no question that the MOS is much more sensitive and reacts far more quickly than the catalytic bead sensor. Calibration plays a factor in the readings being double or just different, as the gas company meter is calibrated for natural gas (or propane) and the responders' is usually calibrated to pentane, and may be set to react differently for natural gas. Some meters use a 0.5 response curve factor for natural gas, which means that the readings will be off by 50 percent, thereby providing an intentional safety factor. The gas company meter does not have this cushion built in. The other factor is that most gas companies calibrate their meters daily, something very few responders do. The next section describes how these various sensors work and how the readings may differ.

NOTE The basic principle of most LEL sensors is that a stream of sampled air passes through the sensor housing, causing a heat increase.

Wheatstone Bridge Sensor

What used to be the most common sensor, a **Wheatstone bridge** sensor (Figure 4-3) is essentially a coiled platinum wire in a heated sensor housing. When the sample gas passes over the "bridge," it heats up that side of the bridge. Platinum is used because it can catalyze oxidation (combustion) reaction at relatively low temperatures and at low concentrations of flammable vapors. If the air sample contains any concentration of a flammable gas, the platinum filament will get hotter and increase its resistance to an electrical current in a nearly direct proportion to the concentration of any flammable content. The change in resistance is compared to a known constant resistance or a second parallel bridge and the difference is converted to a meter reading. Wheatstone bridge is considered to be old technology but is not bad technology, and it has a proven track record. Newer and much better monitors with Wheatstone bridge LEL sensors are actually two separate wire coils in the middle of the sensor; both are heated with the sample gas passing over one of the bridges. The sample gas, if flammable, causes the bridge to heat up, resulting in a difference in the electric resistance in between the two bridges. This difference is reported on the readout of the monitor. These new types are a marked improvement over the old and have corrected many of the problems given in the following list. The advantage of the newer monitors is that they allow for correction of

Figure 4-3 A Wheatstone bridge sensor.

Figure 4-4 A catalytic bead sensor.

ambient temperature and humidity and other factors that cause instabilities to the meter reading.

Wheatstone bridges may have problems in the following areas:

- Low oxygen atmospheres. To get an accurate reading, the minimum amount of oxygen necessary is 16 percent. Less than 16 percent oxygen and the readings will be off and vice versa. If the oxygen is high, then the LEL reading will be high.
- Acute sensor toxins such as lead vapors (from leaded gasoline), sulfur compounds, silicone compounds, and acid gases can corrode or coat the filament, which can cause altered readings.
- Chronic exposure through high levels may saturate the sensor and cause it to be useless for a long period of time until purged and recalibrated.

The old Wheatstone bridge LEL sensors (as well as a many new ones) do not indicate when they go above the LEL. In most cases the LEL sensor would indicate 100 percent of the LEL for a very short time and then bounce back to 0 percent never to rise again, no matter what. The bridge had burned out and would not function. Many of the meters allow the sensor to read up to 100 percent of the LEL, then shut the sensor off, or if so equipped, switch over to the LEL sensor that does percentage by volume methane. This improvement will save many dollars in sensor replacement. The new LEL sensors should last a minimum of four to five years although almost all manufacturers only warranty them for one to two years. The Wheatstone bridge reads on a scale of 0 percent to 100 percent, and starts to indicate at a level of approximately 50 ppm of methane.

Catalytic Bead Sensor

The **catalytic bead sensor** is the most common LEL sensor for emergency response and is essentially the same as the newer Wheatstone bridge technology with some new twists. Instead of twists of wire forming a bridge the catalytic bead sensor uses a bowl-shaped string of metal with a bead of metal in the middle. The metal is typically coated with a catalytic material that helps burn the gas sample off efficiently. The sensing unit, as shown in Figure 4-4 and Figure 4-5 has two sensors, placed in the same fashion as the Wheatstone bridge, one for sampling and the other for reading the change in the sampling bowl. The catalytic bead sensor is much more precise than the Wheatstone bridge and is less susceptible to breakage. With normal use they usually last four to five years. The catalytic bead sensor reads at 0 to 100 percent of the LEL, although some units also read at parts per million levels, usually

Figure 4-5 A catalytic bead sensor.

starting at 50 ppm. One very useful feature on some monitors is the sensor that shuts off at 100 percent of the LEL, which prevents the sensor from becoming saturated or burned out. The catalytic bead also begins to read at about 50 ppm. Some newer catalytic bead sensors are being used to read both percent of LEL and parts per million. The ability to switch measuring modes was formerly reserved for the metal oxide sensor. The ability for the catalytic bead sensor to detect small amounts of a flammable, usually starting at the 50 ppm range, is generally the result of the sample gas being preheated before it enters the catalytic bead chamber. As the gas is heated it reacts differently in the chamber and the chamber will burn off the sample, causing a reading on the monitor. It is important to ask the sales rep what type of sensor the LEL sensing unit is, as there is a difference between a catalytic bead sensor just described and the metal oxide sensor discussed in the next section.

Figure 4-7 A metal oxide sensor.

Metal Oxide Sensor

The MOS or broadband sensor has attracted a lot of negative attention. Most people do not understand how the sensor functions, and since it is a very sensitive sensor, it has a great deal of perceived problems. If used and interpreted correctly, this sensor, shown in Figure 4-6 and Figure 4-7 can provide answers to many response questions.

The MOS is a semiconductor in a sealed unit that has a Wheatstone bridge surrounded by a metal oxide coating. Heater coils provide a constant temperature. When the sample gas passes over the heated bridge, it combines with a pocket of oxygen created from the metal oxide. Anything that enters the sensor housing is burned off, and this reaction causes an electrical change, which produces the reading. Remember that the Wheatstone bridge and catalytic bead sensor only burn off and react to flammable gases, and nothing else. The MOS picks up dust, dirt, other particulates, moisture in the air, flammable gases, combustible gases, or just about any chemical with enough vapor pressure to get it into the air. This sensitivity can present a problem to some people. If anything is on the sensor when the meter is first turned on, it will heat up, trying to burn it off and causing a reading. Most people are not patient enough to let the meter warm up for at least five to ten minutes, let the unit clean itself, and stabilize. In most cases this type of sensor requires frequent calibration, and requires zeroing each time it is turned on. It can be a very frustrating sensor, but provides a lot of valuable information about air quality and potential contaminants. It is so sensitive as compared to the other LEL sensors that it can actually pinpoint tiny leaks in pipes, a valuable benefit in industrial use.

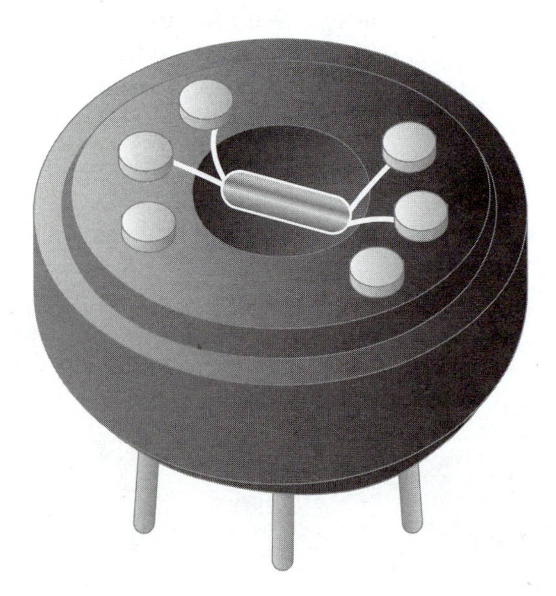

Figure 4-6 A metal oxide sensor.

NOTE The MOS picks up moisture in the air, flammable gases, combustible gases, or just about any chemical with enough vapor pressure to get it into the air.

Most MOS LEL sensors do not provide a readout of the percentage of the LEL as other LEL sensors do, but provide an audible warning or a number within a range. If you spill a tablespoon of baby oil on the table and pass a MOS sensor over it, the sensor will give a reading, although no other LEL sensor would pick this material up. If you take a MOS sensor into a room with 5 percent methane, the reading will be within a range for a flammable gas. Some monitors allow the MOS to read in percentages of the LEL, in addition to a general sensing range of 0 to 50,000 units. The MOS reacts to tiny amounts, which is an outstanding feature of the monitor, as most monitors are not this sensitive. In many cases HAZMAT teams are called to buildings in which tiny amounts of something in the air is causing problems. The MOS detector is very useful in determining whether something is there and pointing to what that unidentified material is. If the department can only afford one combustible gas sensor, it is not recommended that you purchase a MOS unit. Either a Wheatstone bridge or catalytic bead will give adequate performance at a lesser cost. If a response team is looking to further enhance its capabilities, then a MOS is a great addition to its air monitoring capabilities. The MOS sensor usually can be set to read 0 to 100 percent of the LEL or on a 0 to 50,000 unit scale, and starts to detect some gases at levels near 3 to 5 ppm, a great advantage. It is near PID levels for some gases and reacts to many things a PID does not.

NOTE The MOS is so sensitive as compared to the other LEL sensors that it can actually pinpoint tiny leaks in pipes, a valuable benefit and response in industrial use.

NOTE In many cases, HAZMAT teams are called to buildings in which tiny amounts of something in the air is causing problems. The MOS detector is very useful in determining whether something is there and where it may be coming from.

Infrared Sensors

The infrared sensor is new to emergency response and has some unique features that further enhance the HAZMAT team's capabilities. The **infrared sensor,** shown in Figure 4-8 and Figure 4-9 uses a hot wire to produce a broad range of wavelengths, uses a filter to obtain the desired wavelength, and has a detection device on the other side of the sensor housing. The light emitted from the hot wire is split, one through the filter, the other to the detection device to be used as a reference source. When a hydrocarbon gas is sent into the sample chamber, the gas molecules absorb some of the infrared light that does not reach the detection device. The amount of infrared light reaching the detection device is compared to the reference source and if there is a difference the meter outputs a reading.

NOTE The big advantage of infrared over other LEL sensors is that it does not require oxygen to function; it can read flammables in oxygen-deficient atmospheres.

The big advantage of infrared over other LEL sensors is that it does not require oxygen to function; it can read flammables in oxygen-deficient atmospheres. One common situation in which this capability this is useful is in ship fires or other situations where a response tactic is to inert the atmosphere. Inerting agents such as nitrogen are commonly used in cargo hold fires, or situations in which materials are reacting in a hold of a ship. When atmospheres are inerted, usually with nitrogen, the goal is

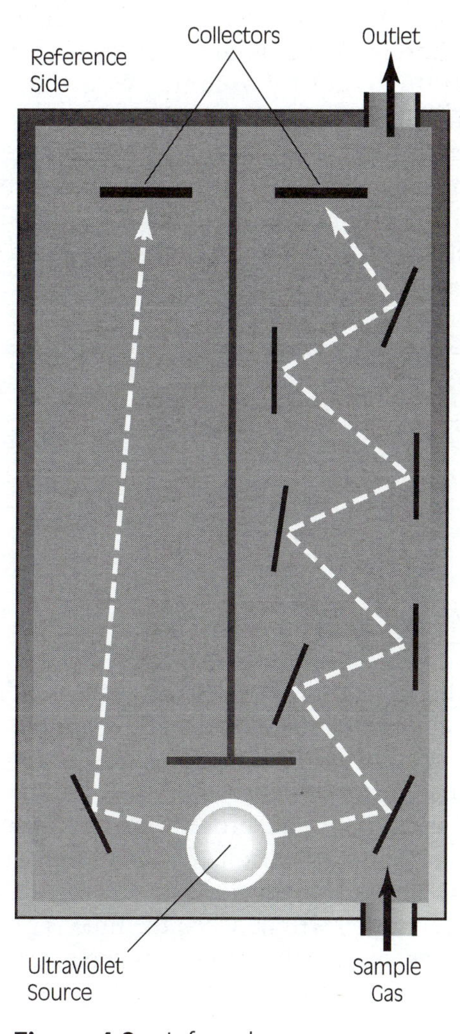

Figure 4-8 Infrared sensor.

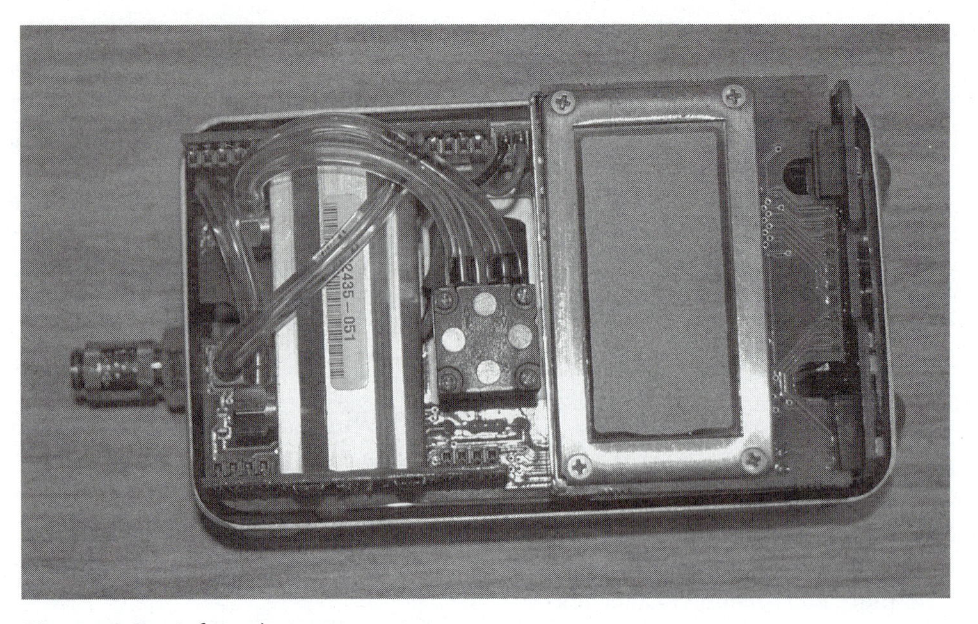

Figure 4-9 Infrared sensor.

to have a 0 percent oxygen atmosphere. The easiest way to determine if this tactic has worked is to monitor for flammable gases (usually hydrogen), but the low oxygen level presents some stumbling blocks for monitoring. Most infrared meters also measure up to 100 percent by volume. When the levels exceed 100 percent of the LEL, the meter then switches to percent by volume, which means it can read upward from the LEL to 100 percent by volume. The device also is not affected by temperature, nor is it easily poisoned by high exposures. Disadvantages include cost and many cross-sensitivities.

NOTE Most infrared meters also measure up to 100 percent by volume. When the levels exceed 100 percent of the LEL, the meter then switches to percent by volume.

LINEARITY

When discussing the various LEL sensors, the subject of linearity arises, and how it applies to response work. The Wheatstone bridge, catalytic bead, and infrared sensor are linear sensors. By being linear they can provide estimations of an anticipated concentration. By knowing what the concentration is at one point, you can calculate what the concentration will be when the time is extended. Record one spot on the graph at 0 percent, which is one known point, and then record the next point, which is another known. When you plot these out and extend the line, you can anticipate future concentrations. Another way of describing linearity is that the readings are accurate to the actual concentration across some portion of the scale, or the whole scale before the reading deviates between actual and measured amounts. A MOS is not linear, and no matter how many points on the graph you plot, the readings will be all over the place with no rhyme or reason. Even in known concentrations where the level does not change, the reported readings will be scattered. A MOS is valuable as it detects tiny amounts of things in the air, but is irritating as it is nonlinear.

SUMMARY

The ability to detect the fire hazard is important because flammables can cause problems over a large area and you could be in it. Some flammables are also toxic, so they present a double risk. Flammables and combustibles are the leading category of materials that records show are released in this country, and they are our most frequent responses. Knowing the type of LEL sensor in your instrument is key to solving many air monitor issues. Knowing the various types of sensors that exist make you an educated consumer and allow you to purchase instruments that best suit your needs.

KEY TERMS

Catalytic bead sensor The most common type of LEL sensor; uses two heated beads of metal to detect the presence of flammable gases.

CGI See *Combustible gas indicator.*

Combustible gas indicator A monitor designed to measure the relative flammability of gases and to determine the percent of the lower explosive limit. Also known as an LEL monitor.

Infrared sensor A type of LEL sensor that uses infrared light to detect flammable gases.

LEL meter The best name for a meter that is used to detect flammable gases.

LEL sensor A sensor designed to look for flammable gases; can be of four designs, Wheatstone bridge, catalytic bead, metal oxide, and infrared.

Metal oxide sensor A form of LEL sensor.

MOS See *Metal oxide sensor.*

Tote A portable container that holds solids, liquids, and gases. Liquid totes hold 300 to 500 gallons of various products—an increasingly common method of chemical storage and shipping.

Wheatstone bridge A form of LEL sensor.

TOXIC GAS SENSORS

- Introduction
- Toxic Gas Sensing Technology
- Carbon Monoxide Incidents
- Summary
- Key Terms

INTRODUCTION

This chapter discusses the sensors most commonly used in three- four-, or five-gas units (LEL, O_2, Toxic/Toxic/Toxic) (see Figure 5-1). These are often used to measure carbon monoxide (CO), hydrogen sulfide (H_2S), and chlorine (Cl_2). **Toxic gas sensors** are available in a variety of materials, however, and include the capability to detect ammonia, sulfur dioxide, hydrogen chloride, hydrogen cyanide, nitrogen dioxide, and many others. The most common unit sold today by far is a four-gas unit that measures LEL, O_2, CO, and H_2S; its popularity is a direct result of the confined space regulation issued by OSHA. In the confined space regulation OSHA requires monitoring for oxygen levels, flammable gases, and at least two toxic gases. Carbon monoxide and hydrogen sulfide are the most common gases found in confined space entries, particularly sewers and manholes.

NOTE The most common unit sold today by far is a four-gas unit that measures LEL, O_2, CO, and H_2S; its popularity is a direct result of the confined space regulation issued by OSHA.

Figure 5-1 A typical gas meter that measures for toxic gases using electrochemical sensors.

Many response teams consider some of the just-mentioned toxic sensors such as chlorine and ammonia as their fourth or fifth gas. This decision may not be wise economically, as these sensors are costly to operate. Typically HAZMAT teams use colorimetric tubes to monitor for these other gases. Colorimetric tubes cost about $50 per set of ten tests and have a shelf life of two years. Response teams should determine their department's annual cost of doing this colorimetric sampling and compare it to the upkeep of the toxic sensor. The average cost for a CO or H_2S sensor is about $150, but the cost of a toxic sensor for chlorine or ammonia is $400 to $500. These sensors usually last less than a year, the calibration gas is $300 to $400, and has a six-month shelf life. Colorimetric sampling is discussed in Chapter 7. The other problem is that if used in more than 20 ppm chlorine, the entire unit could be ruined. For teams that respond to ammonia or chlorine leaks on a regular basis and who have the financial resources for only one unit, a single gas unit is recommended for purchase for that response. Having just a single gas unit limits the potential damage to that unit. Keep in mind that ammonia is a constituent of most heavy-duty refrigeration systems used for food, beverage, and other cold storage, thus may be the most-needed monitor. Water treatment systems (drinking or wastewater) would likely have chlorine.

SAFETY Carbon monoxide and hydrogen sulfide are the most common gases found in confined space entries, particularly sewers and manholes.

TOXIC GAS SENSING TECHNOLOGY

Most toxic sensors such as shown in Figure 5-2 are **electrochemical sensors** with two or more electrodes and a chemical mixture sealed in a sensor housing. The gases pass over the sensor, causing a chemical reaction within the sensor and an electrical charge, which causes a readout to be displayed. All toxic sensors display in parts per million. Some toxic sensors use metal oxide technology and react in the same fashion as an LEL MOS, and have the same MOS issues.

For some of the problems related to toxic sensors, we must refer to the chemistry of hazardous materials and the periodic table, which plays an important factor when dealing with toxic sensors. The periodic table is set up with families in vertical columns in which all of the elements in a column have similar characteristics. Using the chlorine sensor as an example, chlorine belongs to the halogen family. The halogen family also has fluorine (gas), bromine (liquid with high vapor pressure), iodine (solid with a high vapor pressure), and astatine (solid). Each of these materials has similar characteristics, and the sensor reacts to vapors from them similarly. Although the sensor cannot determine the difference between these materials, it reacts and provides a reading. Responders who do not know what they are sampling could misinterpret.

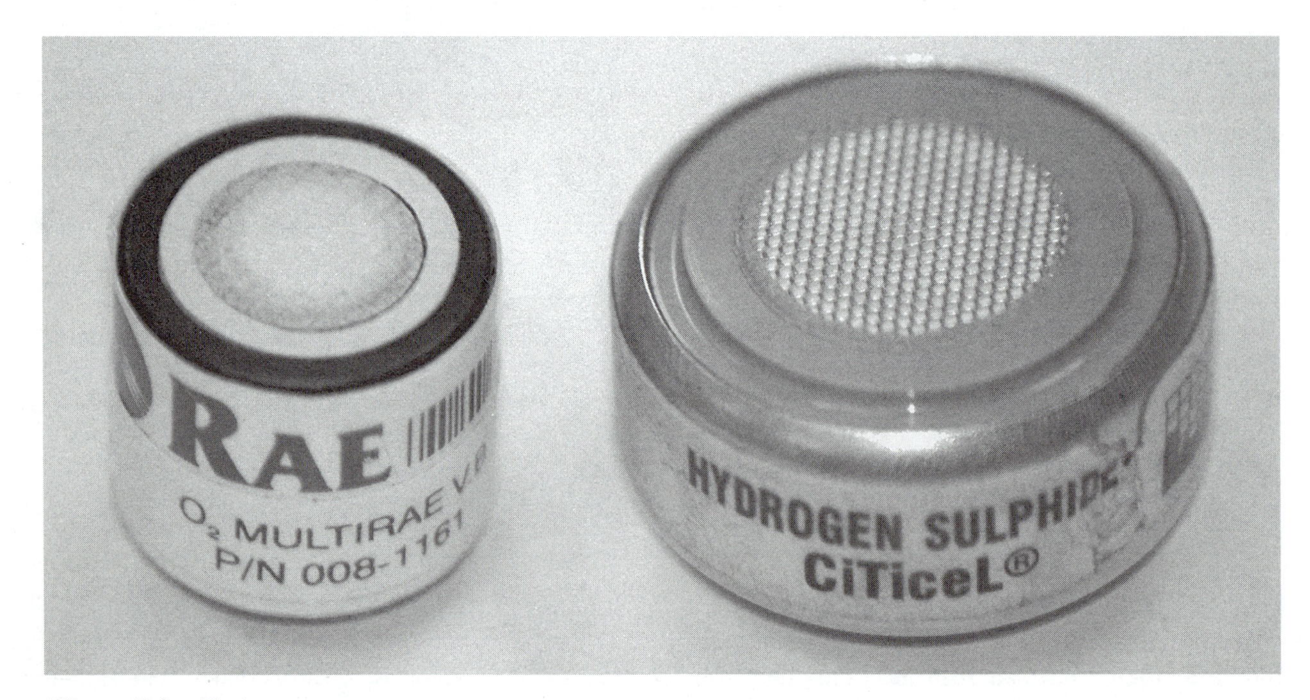

Figure 5-2 Toxic gas sensors.

TABLE 5-1

Interfering Gases to CO and H_2S	
Acetylene	Isobutylene
Dimethyl sulfide	Turpentine
Ethyl alcohol	Ethyl sulfide
Ethylene	Nitrogen dioxide
Ammonia	Methyl alcohol
Phosphine	Isopropyl alcohol
Methyl sulfide	Hydrogen cyanide
Sulfur dioxide	Hydrogen
Propane	Carbon disulfide
Mercaptans	Nitric oxide
Hydrogen sulfide	

An additional complication is that other chemicals can cause **interference** with the sensor. Interfering gases are listed in Table 5-1 and are also described in the case study Corrosive Atmospheres and Salespeople. An example of such an interfering gas for the CO sensor is H_2S. Most CO/H_2S sensors are filtered or manufactured to eliminate the interference but some are not, so buyer beware. At times, a simple charcoal filter over the sensor makes a great difference in obtaining accurate readings. In some cases, high flammable gas readings bleed over to the CO and H_2S sensors causing false readings, usually when the LEL sensor is reading about 40 percent or higher. Before you buy a meter, test several units, and expose them to a variety of materials to see how they react. The CO and H_2S sensors should not react to a flammable liquid such as acetone. If one or both of the toxic sensors does react to the acetone, the sensor(s) are not properly filtered and a different meter should be considered. The toxic sensors and the oxygen sensor have a variety of shelf lives, with many lasting twelve to eighteen months. The amount of time they last depends on the kind of atmospheres to which they are exposed. The sensors have a maximum exposure limit; surpassing this limit kills the sensor. Usually low levels over a period of time do not dramatically change the life of the sensor; high exposures shorten the life of the sensor. When other gases that should not affect a sensor do affect it, its life is shortened.

CASE STUDY

Corrosive Atmospheres and Salespeople—Corrosive atmospheres and salespeople are not related, but both stories relate to toxic gas monitoring. Teams responding one evening to a reported drum leak in a warehouse had little additional information. The identity of the leaking material would take some recon. The first responders entered the building to check conditions, taking with them a four-gas air monitor (O_2, LEL, H_2S, and CO), which was a good idea. They reported that they were getting both CO and H_2S readings of about 70 to 90 ppm and were advised by the HAZMAT team to leave the building as quickly as they could. It would be unusual to find both CO and H_2S in the same place at the same time. Also there were no reported odors associated with the response, and H_2S stinks of rotten eggs. As it turns out, there was a leaking drum emitting corrosive vapors, which was causing the sensors to read.

The salesperson portion of this case study relates to my knowledge of air monitoring. I am not afraid to ask questions, so when a monitor does not do what I think it should, I ask why. We also use many monitors in training situations, and we like to use live chemicals to show how various monitors respond. One brand of monitor, when placed in proximity to a flammable liquid, provided high readings on the CO sensor, which was unusual and caused problems in explaining the reason to the students. The CO sensor was becoming saturated and reacting, causing a reading. None of our other monitors did that, so I took the cover off of a couple and compared the CO sensors. All of the other sensors had a filter on top of them, but the reacting monitor did not. I discussed this issue with the manufacturer's sales rep who defended the product but stated that he would get back to me. The company sent an engineer who stated that their meter was better than any other and that the others were malfunctioning. After discussing the calibration issues and how we were using these meters for training, he still maintained that his were better. When asked how CO is emitted from acetone in a jar, his answer was that this was just training, it would not happen in real life. At the spill pad we spilled some acetone and showed that his meter was reading CO and none of the others were. A week later I got a new pump cover for the meter that had a filter built in it, just like the others. The moral of the story is always try the meter, especially in situations in which you will be using them, and trust the meter; they do not lie.

NOTE The CO and H_2S sensors should not react to a flammable liquid such as acetone. If one or both of the toxic sensors does react to the acetone, the sensors are not properly filtered.

Smart sensors have some promise in the future as the technology advances. The thought process behind smart sensors is that departments keep a variety of toxic sensors in their inventory, and when they need to sample a particular gas, they just plug in the smart sensor. The smart sensor has a computer chip imbedded in the sensor that tracks calibration and other gas specific information the meter requires. Although one would think that this system would be perfect for emergency response, it still needs some improvement. Each time you swap out the sensor and replace it requires calibration, and you need calibration gas for each sensor. Some models do not require calibration each time the sensor is switched, but regular calibration is required, a costly item for toxic sensors. The idea behind smart sensors is a good one, and when sensors are developed that are factory calibrated and do not require any user calibration, it would be a great benefit to smart sensor technology.

CARBON MONOXIDE INCIDENTS

Responses to carbon monoxide detector alarms are increasing for the fire service. In 1995 the city of Chicago ran several thousand carbon monoxide detector alarms in one day due to an inversion, which kept the smog, pollution, and carbon monoxide at a low elevation within the city. As carbon monoxide is colorless, odorless, and very toxic, it is important that first responders understand the characteristics of carbon monoxide, and how home detectors work. As with other chemicals, CO can be an acute or chronic toxicity hazard. It is only acutely toxic at levels in excess of 100 ppm. At levels less than 100 ppm the hazard comes from a chronic exposure. A chronic exposure to CO can be hazardous because many people are exposed to CO in the home. The time people spend in their homes constitutes a considerable exposure.

Carbon monoxide can only be detected by a specialized detector. In extremely high concentrations it can be explosive. Exposure to CO causes flulike symptoms—headache, nausea, dizziness, confusion, and irritability. Depending on the length, exposure to high levels can cause vomiting, chest pain, shortness of breath, loss of consciousness, brain damage, and death.

SAFETY CO poisoning signs and symptoms may be delayed for twenty-four to seventy-two hours. Levels over 100 ppm are extremely dangerous and the residents should be medically evaluated. Fatalities have occurred in the developing fetus at 9 ppm.

When exposed to levels of CO in a house, the residents may not exhibit any signs or symptoms. If CO is detected, they should seek medical attention because CO poisoning signs and symptoms may be delayed for twenty-four to seventy-two hours. Levels over 100 ppm are extremely dangerous and the residents should be medically evaluated. Fatalities have occurred in the developing fetus at 9 ppm. Monitoring with a CO detector is essential to determine the possible exposure to CO. People who may only show minor effects of CO poisoning and who would normally be transported to the closest hospital need to have their residences monitored. If high levels are found using a monitor, then the **hyperbaric chamber** may be the best treatment and should not be delayed. Many times a Pulse-Ox (oxygen saturation monitor) will be used incorrectly to determine the O_2 level in a patient. Patients who have been exposed to CO will cause a Pulse-Ox to read 100 percent as the monitor reads the oxygen molecule in CO as being O_2. The elderly, children, or women who are pregnant are especially susceptible to CO and may have had a serious exposure without showing any effects.

If the first-arriving units do not have a CO monitor and victims may be remaining in the residence, personnel are to don functioning SCBA when searching the residence. After determining no victims are present, crews are to ensure that the house is closed up and then wait outside for a CO monitor. Do not enter an area with an activated CO detector without the use of SCBA. If crews find levels that exceed 35 ppm according to the monitor, they should use SCBA to continue the investigation. Crews should be suspicious when responding to reports of an unconscious person, or of "several persons down." They should not enter an area without SCBA if it is possible that CO (or other toxic gases) may be present. An air monitor will ensure responder safety, from the gases for which it samples.

It is possible that people may be found unconscious due to a natural gas leak, but this is very unlikely. Natural gas, which has a distinctive odorant added to it, is nontoxic and only asphyxiates a

HOME CO SENSOR TYPES

Biomimetic is a gel-like material designed to operate in the same fashion as the human body when exposed to CO. This sensor is prone to false alarms, as it can never reset itself, unless it is placed in an environment free of CO, which in most homes is impossible. The sensor may need twenty-four to forty-eight hours to clear itself after an exposure to CO. The actual concentration of CO at the time the detector sounds may be low, but the exposure may have been enough to send the detector over the alarm threshold. If responding to an incident in which one of these detectors has activated, it should be placed in a CO-free environment until it clears. In the meantime, residents should rely on an alternate CO monitor. They should not be unprotected.

The metal oxide detector is similar to sensors used in combustible gas detectors but is designed to read carbon monoxide. How successful this design is in only reading CO is subject to debate. Although the metal oxide sensor is superior to the biomimetic sensor, it has some cross sensitivities and will react to other gases. Although it is hoped that the responders would be using a three-to-five-gas detection device to check a home, it is possible for this type of sensor to alarm for propane. Responders using only a CO instrument may find themselves walking into a flammable atmosphere. Metal oxide detectors can usually be identified by the use of a power cord as the sensor requires a lot of energy. In most cases these sensors provide a digital readout. Once activated this sensor needs some time to clear itself, usually less than twenty-four hours.

An electrochemical sensor, which may also be referred to as instant detection and response (IDR), is the same type of electrochemical sensor that is in a three-, four-, or five-gas instrument. It has a sensor housing with two charged poles in a chemical slurry. When CO goes across the sensor, it causes a chemical reaction that changes the resistance within the housing. If the amount is high enough, it will cause an alarm. This sensor provides an instant reading of CO, and does not require a CO buildup to activate. It has an internal mechanism that checks the sensor to make sure it is functioning, which is a unique feature. Out of the three types of residential detectors, based on sensor technology the electrochemical monitor would provide the best sensing capability.

person by pushing oxygen out of an area. The only sign of this exposure is unconsciousness or death; any of the flulike symptoms are due to CO poisoning. If the level of natural gas is high enough to cause unconsciousness, then a very severe explosion hazard is present, and in reality an explosion would be imminent.

It is possible that standard fire department air monitors will not pick up any CO, because the CO detectors purchased for homes are made to detect small amounts of CO over a long period of time, and fire department detectors are instant reading that only pick up 1 ppm or more. The fact that responders may not pick up any readings does not mean the residents' detector is defective. Some of these factors that account for this discrepancy are low amounts of CO or a momentary high level that activated the alarm but dissipated prior to responders' arrival. The amount of time the residence is open dramatically affects readings. Crews are reminded to keep the residence as closed as possible so that the air monitor has a chance to monitor the level of CO. As a reminder, any time you respond to unidentified odors or sick building calls, it is important to remove any people from the building and keep it closed up. As the amounts of toxic gases in sick building incidents are usually small, keeping them contained is very important. You cannot treat patients of toxic gas exposure unless you have a clue as to the source. For the patients' long-term health and your own, you need to obtain quick reliable gas samples.

It some parts of the country CO detectors are required just as smoke detectors are. Performance of these detectors varies from brand to brand. The three basic sensing technologies are **biomimetic,** metal oxide, and electrochemical, each with advantages and disadvantages. Location, weather conditions, and the type of sensor determine the types of readings that can be expected from a particular brand of detector.

Common sources of CO include:

- Furnaces (oil and gas)
- Hot water heaters (oil and gas)
- Fireplaces (wood, coal, and gas)
- Kerosene heaters (or other fueled heaters)
- Gasoline engines running inside basements or garages
- Barbecue grills burning near the residence (garage or porch)
- Faulty flues or exhaust pipes

SUMMARY

The use of toxic sensors partially protects against the toxic hazard, as the toxic sensors are specific to a few gases. Responding to CO events is fairly commonplace, and is a response that you should be competent in handling. The response to CO is the most common event in which monitoring is used in combination with patient rescue and treatment. When making toxic gas sensor purchase decisions, keep in mind the costs of operating a meter, and calculate how often the electronic meter would be used.

KEY TERMS

Biomimetic A type of CO sensor used in home detectors.

Electrochemical sensor A sensor that has a chemical gel substance that reacts to the intended gas and provides a reading on the monitor.

Hyperbaric chamber A pressurized chamber that provides large amounts of oxygen to treat inhalation injuries, diving injuries, and other medical conditions.

Interferences Gases picked up by the sensor but not intended to be read by the sensor.

Smart sensor Sensor that has a computer chip on it that allows the switching of a variety of sensors within an instrument.

Toxic gas sensor Device for the detection of toxic gases. Common toxic gas sensors are for CO, H_2S, ammonia, and chlorine.

IONIZING DETECTION UNITS

INTRODUCTION

Two major detectors of potential toxic risk, the **photoionization detector (PID)** and the **flame ionization detector (FID)** comprise the ionizing detection type. Other ionization detectors are discussed in Chapter 11. Ionization is the splitting of a vaporized molecule into separate electrical charges, positive and negative. When a vaporized molecule is ionized, it is split into both types of charges, and the detection device measures the electrical activity and provides a reading. The method of ionization may vary but the end result is the same: The vapor is separated and the resulting change in electrical activity is measured against that of a known gas, which is the calibration gas. Both the PID and the FID use ionization to detect a variety of gases, but the mechanism to complete that ionization differs. As can be seen in Table 6-2 the LEL sensor does not detect at a low enough level to protect responders against toxic risks. For HAZMAT teams, a PID is an essential device, one whose value cannot be overstated. Departments without a PID are taking a serious risk as they do not have an easy method of detecting common potentially toxic materials.

NOTE Both the PID and the FID use ionization to detect a variety of gases. These detectors identify the potential toxic risk that chemicals present.

SAFETY Departments without a PID are taking a serious risk, as they do not have an easy method of detecting common potentially toxic materials.

PHOTOIONIZATION DETECTORS

Sometimes referred to as a total vapor survey instrument, a PID (see Figure 6-1) can detect organic and some inorganic gases, including ammonia, arsine, phosphine, hydrogen sulfide, bromine, and iodine. Because of its ability to detect a wide variety of gases in small amounts, it is an essential tool of response teams. The PID does not indicate what materials are present, it just identifies that something is in the air. Used as a general survey instrument, the PID can alert responders to potential areas of concern and possible leaks/contamination. The original PIDs were designed for the petroleum refining and storage industry and are widely used during underground storage tank (UST) removals. Because of their sensitive nature, they can detect small amounts of hydrocarbons in the soil. Sick building calls are on the increase, and the PID is a valuable tool in identifying possible hot spots within the building. PIDs even have a valuable use in possible terrorism events because all the chemical agents can be detected by a PID.

Figure 6-1 A typical photoionization detector with the additional benefit of a four-gas detector combination. In addition to the PID, this unit measures oxygen, LEL, carbon monoxide, and hydrogen sulfide.

NOTE The PID does not indicate what materials are present, it just identifies that something is in the air.

The biggest advantage that a PID has is its sensitivity as it starts to read at 0.1 ppm. PID monitors can read up to 2000 ppm and several monitors read to 10,000 ppm. This reading in parts per million is for the calibrated gas, which is usually isobutylene. When the gas is unknown, we call the reading *meter units*. A reading of 135 on a PID that reads 0.1–2,000 ppm for an unidentified gas would be called 135 meter units, which means the meter moved 135 units out of a scale of 2000. RAE systems has a PPB PID that reads into the parts per billion, a very sensitive instrument. Chemicals with a **permissible exposure limit (PEL)** or **threshold limit value (TLV)** of less than 500 ppm are considered toxic, so the PID is useful in identifying these levels. The problem with using the LEL sensors with the exception of the MOS is the fact that they require high levels to begin to read. Except for the MOS, LEL sensors do not begin reading unless the levels exceed 50 ppm. We can convert the readings found on an LEL monitor to parts per million by using the formula found in Table 1-8. Some conversions are provided in Table 6-1.

TABLE 6-1

Conversions from Percentage of Volume to ppm

Percentage of Volume	Equivalent ppm
1	10,000
2	20,000
5	50,000
10	100, 000
20	200,000
30	300,000
40	400,000
50	500,000
60	600,000
70	700,000
80	800,000
90	900,000
100	1,000,000

It is easy to see that one could be in significant trouble in regards to toxicity, when an LEL sensor would not even indicate a problem. The PID is used to look for small things in air, and LEL sensors are for the larger problems, most specifically the fire risk.

Table 6-2 presents another example as to how it is easy to be fooled into thinking the atmosphere is safe. Suppose that you are responding to a reported odor of gas in an apartment building. Upon arrival you take your four-gas air monitor into the building. When responding to a reported gas odor, at what point do you don your facepiece? What meter reading (LEL sensor) would make you put your facepiece on? Do citizens always know natural gas from other chemical odors? Think about these answers before you read the table.

The table shows that at 0.8 percent of the LEL, there are 141 ppm of phenol in the air. The LEL sensor at 0.8 percent would be reading 0, as it requires at least 1 percent to indicate. At 1.6 percent the

TABLE 6-2

Toxicity Versus Fire Risk

Known Information	Values
Actual Spilled Chemical	Phenol
LEL	1.8%
OSHA PEL	5 ppm
NIOSH REL	5 ppm
IDLH	250 ppm

Reading from Catalytic Bead LEL Sensor Calibrated to Phenol (in percent)	Parts Per Million in Air Equivalent
100	18,000
50	9,000
25	4,500
12.5	2,250
6.7	1,125
3.3	563
1.6	282
0.8	141

meter would actually read 1 as it only reads whole numbers. If you just used the LEL sensor you could be in an environment that is more than half the IDLH without the monitor indicating any level. Many responders use 10 percent or 25 percent of the LEL in which to put their facepieces on or to take action, which would be 1688 and 4500 ppm respectively. At these levels there is a severe toxic risk to an unprotected responder. However, a PID with an 11.7 electron volt (eV) lamp would pick up this material at 0.1 ppm, a level that would offer far greater safety.

The PID uses an ultraviolet (UV) lamp to ionize any contaminants in the air, which splits up the electrical charges in the molecule. These positive and negative charges can then be read by the detector in the sensor housing. In order to be read by a PID, the vapor or gas to be sampled must be able to be ionized, which is called **ionization potential (IP).** The unit of measurement of an IP is electron volts (eV). There are various types of UV lamps available; the most common are a 10.2, 10.6, and 11.7 eV. In order to read a vapor or a gas with a PID, the gas must have an IP less than the eV rating on the lamp. For example, in order to sample benzene, which has an IP of 9.2 eV, we must use a lamp of at least 9.2 eV or above. The standard 10.6 eV lamp would be acceptable to read benzene. In most cases, gases that have IPs above the lamp strength cannot be read, although there is some carryover. The IP of water is in excess of 12, but high levels of water in the air (humidity and fog) will cause a PID to read some of the water vapor, as the water has a tendency to short out the electrical contacts in the sampling chamber. Some common IPs are provided in Table 6-3. Vapors in excess of the lamp strength will read in some fashion, but it is nowhere near an accurate reading. Keep in mind that the PID is calibrated to a specific material (usually to benzene using isobutylene as the calibrating gas) so relative response curve factors apply here as they did with LEL sensors. The LEL sensors have low response factors, but the PIDs have some large numbers for response factors. PIDs are able to pick up readings from toxic substances, but also can detect baby oil, motor oil, gasoline, and many other hydrocarbons. Many liquid pesticides are 0.5 percent to 50 percent solution mixed with xylene, trimethyl benzene, and emulsifiers that are easily detected by the PID. Most drum dumps spill waste oil, fuel oil and the like so the PID is a valuable resource in protecting responders and the public from toxic materials.

One confusing issue with the PIDs is what is considered to be a toxic reading for unidentified gases when using a PID. The general rule of thumb for an occupancy that has chemicals in use is that a reading of less than 50 ppm is acceptable, and for a PPB PID a reading of less than 500 ppb is normal. There is no easy answer when trying to determine what constitutes a toxic environment when using a PID,

TABLE 6-3

Ionization Potentials of Common Materials

Chemical	Ionization Potential (eV)
Acetone	9.69
Acetylene	11.4
Ammonia	10.18
Ethyl ether	9.53
Formaldehyde	10.88
Freon 112	11.30
Hydrazine	8.93
Hydrogen cyanide	13.60
Hydrogen fluoride	15.98
Hydrogen peroxide	10.54
Hydrogen sulfide	10.46
Methane	13.0
Methyl alcohol	10.84
Methyl ethyl ketone	9.54
Nitrous oxide	12.89
Phenol	8.5
Phosgene	11.55
Propane	11.07
Sulfur dioxide	12.30
Toluene	8.82
Triethylamine	7.5
Xylene	8.56

as many other factors come into play in answering that question.

The two primary considerations are location or occupancy and biological indicators. There are certain occupancies, such as print shops, gas stations, paint, and auto parts stores to name a few, in which a PID will read the vapors in the air. The determination in whether that atmosphere is toxic is based on the predominant material in the air and what that particular chemical's PEL is. It can be anticipated that in a paint store where throughout the day containers have been opened and spills are likely to have occurred, vapors may escape their containers. This mixture in the air would be considered normal and probably would not exceed the PEL. You could anticipate readings of 20 to 300 meter units in this type of occupancy. However, in a home, other than in the garage or the basement, there should not be any toxic vapors in the air, so the PID should read close to the background level. If the PID is used in the garage it could be anticipated that some readings might be obtained in the area where chemicals may be stored, but you would have to be fairly close to the containers, and the readings should be low. As another example, in a bedroom nothing should cause the meter to go above background. If you do find readings, then there is a problem in that house that needs to be corrected. That does not mean that the problem is toxic or life threatening, only that there are vapors in the air that should not be there.

The biological indicators are also crucial to an investigation. If the occupants of the building have any signs or symptoms and you are reasonably certain that they are real, then any reading above background on a PID should be taken into account. Such a reading means is that there is something in the air in tiny amounts that is causing people problems. Encountering dead people and getting a PID reading of 1 meter unit means that something extremely toxic is in the air, in tiny amounts, and at those levels has killed some people. The best way to really learn how to use a PID and to be able to interpret any readings is to use the PID all the time. When doing inspections or facility tours, take the PID with you. Use the meter in "normal" situations to learn what sets it off, and what anticipated readings you may find.

NOTE The best way to really learn how to use a PID and to be able to interpret any readings is to use the PID all the time.

Following is a list of some problems with PIDs:

- The biggest factor in the use of a PID is that humidity plays a role in the reading. Some manufacturers have automatic humidity adjustments for their instruments.
- The lamps are affected by dirt and dust and require cleaning. Environments that have diesel exhaust and other particulate matter such as mown grass may affect the meter. Salt water, or hard water environments may affect the lamp as well.
- Higher levels of methane (natural gas, swamp gas, landfill gas) may suppress some of the ionization potential of the lamp. Use an LEL monitor to read the LEL because a PID does not read methane (IP of 13.0).
- The PID cannot separate out mixed gases, and a mixture can present identification problems. Using various strength lamps can help separate out a gas mixture, but to do so takes extensive time and thought.
- With some PIDs you may need at least 10 percent oxygen in the air for the PID to function.

CASE STUDY

The HAZMAT unit was called for mutual aid to an adjacent county to assist with a fire in an apartment building that had been burning for some time, but had no known hazardous materials. What was known was that two firefighters were in critical condition at Shock Trauma, one in respiratory arrest. They were being sent to the hyberbaric chamber for treatment of some unknown respiratory problem. When we arrived it was determined that about thirty other firefighters were experiencing some problems as well. The fire was still burning but there was no report of any materials that had been found that may have been the source. Some air monitoring on the scene was done without much success. On a whim we decided to check the air bottle of one of the downed firefighters. When the PID was put into the air stream it provided a hit. We set up some apparatus to test the air in the SCBA bottles. We found that the air in the SCBA was providing readings on the PID, and that in the colorimetric testing we found some questionable contaminants and carbon monoxide in excess of 70 ppm. These problems were found in many of the bottles we had confiscated. At 3 A.M. we contacted our local lab, and with split samples we headed to the lab. We would drop the other samples off at the state lab in the morning. That morning the lab came back with the test results from a gas chromatograph/mass spectrometer and reported that the air had components of gasoline in it. We identified the source as a compressor unit that had compressed on the scene, the engine was not running well, and there were some concerns with the location of the air intake hose. The PID followed up by colorimetrics was essential to solving this problem. The firefighters recovered and were released a few days later.

The RAE systems PID for an example does not require any oxygen to function and has been tested in oxygen-deficient atmospheres.

FLAME IONIZATION DETECTORS

One manufacturer refers to its product as an Organic Vapor Analyzer (OVA), but this instrument is an FID with the added benefit of a **gas chromatograph (GC).** This class of instruments used to be referred to primarily as OVAs but FID is the current term used to describe them. The FID works on a principle similar to the PID but has some capabilities beyond a PID. Instead of using UV light to ionize any gases that may be present, the FID uses a hydrogen flame to complete the ionization process. The use of hydrogen to provide the flame can be considered a disadvantage because the FID only detects organic materials in the air, whereas the PID detects both organic and inorganic vapors. Using both of these tools is the start of the detection process. If the PID detects the material and the FID does not, then the material is inorganic. If they both detect the material, then it is organic. The biggest advantage of the FID over the PID is that it can be considered a more sensitive device. But this sensitivity can also be an extreme disadvantage. Because it can read so many varied gases, it generally gives some type of reading, which can be misinterpreted. Taking a PID up to a closed sample jar that contains a flammable liquid will generally not result in any reading above background. When using an FID though, vapors escaping the closed container can be detected. When a PID ionizes a gas sample, it is a one-time chance for a successful ionization; the FID, however, continues to try to ionize the sample, allowing for even greater results. Because the FID uses a flame to ionize the gas and the resulting release of carbon atoms is a repeatable event, the FID is a more accurate device than the PID. The FID has the ability to read methane gas, while the PID does not read methane and high levels of methane will depress any PID readings for other materials that may be present. The FID reads methane very well, actually down to .5 ppm. As with other instruments the FID needs good oxygen to work, and high wind conditions may cause the flame problems. The hydrogen cylinder on the unit may need to be refilled during long duration sampling and doing so may be cumbersome. The FID does very well at detecting a small amount of toxic material in the air, but does not do well on the high end readings, except for units that auto scale up to 50,000 ppm. When dealing with concentrated vapors such as those in the head space of a drum there is a tendency for the FID to flame out. As was suggested with the PID, the best way to get familiar with this device is to use it routinely during nonemergency situations to see what it reacts to and what it does not.

GAS CHROMATOGRAPHY

Gas chromatography, or GC, is a possible next step for response teams to take after the use of an FID. Although not in common use today, due to increasing concerns for terrorism it is likely to be in the future. Several companies produce field instruments that have some application in emergency response. The field instruments are usually coupled with a PID or an FID to complete the sample analysis. The GC uses a heated column through which a sample is passed in the presence of an inert gas such as helium. Each component of the sample has an attraction for the material that makes up the column, and has a specific holdup and travel time within the column. When the material exits the column, the PID or FID detector reacts to each component in proportion to its relative concentration in the sample with respect to its travel time. A graphical picture of travel times and relative concentration is compared in the library, and a match of components can hopefully be made. With multiple gases present, each will "boil off" at differing temperatures causing an individual spike on a roll of paper (much like the heart beats on a heart monitor). Each chemical has a *retention time* and produces a spike at a specific time interval, which identifies the particular component. The length and width of this spike (shown in Figure 6-2) can be used to determine the relative amount of each component material when the printout is compared against known samples. The sampling time varies depending on the volatility of the whole sample, but can often be completed in ten to twenty minutes. GC is one of the last resorts when dealing with the otherwise unidentified, but the lab must have a comparison sample of the gas or liquid you want them to analyze for them to even begin to hazard a guess. A GC cannot detect water, so if the sample contains water, the other components will be overstated unless the water content is independently measured and factored into the analysis calculations. Field GCs are available but they are not readily used by response teams due to their highly technical nature and high cost. Although this technology has changed a great deal in the past few years it has become much easier than it had been. There still is a need for much training to master the use of this device, and the unit needs to be run and calibrated at least weekly.

MASS SPECTROMETRY

The **mass spectrometer (MS)** is usually coupled with the GC, and it is really a seamless operation with the two technologies. The MS of a GC/MS is the actual identifying portion of the device. The GC separates the molecules as discussed previously and then the MS portion provides the identification. To accomplish this identification, the MS actually measures the relative mass of the molecular fragments and compares them to the mass of a comprehensive list of materials in its library, as each molecular fragment has a different weight. In order to identify a material, the footprints must be in the library. This library is the most important part of the GC/MS, and the more extensive the library, the more successful you will be at identifying an unknown substance. If your team cannot afford a GC/MS then it is recommended that you locate a local lab that will run samples for you. Some local police department crime labs have GC/MS capabil-

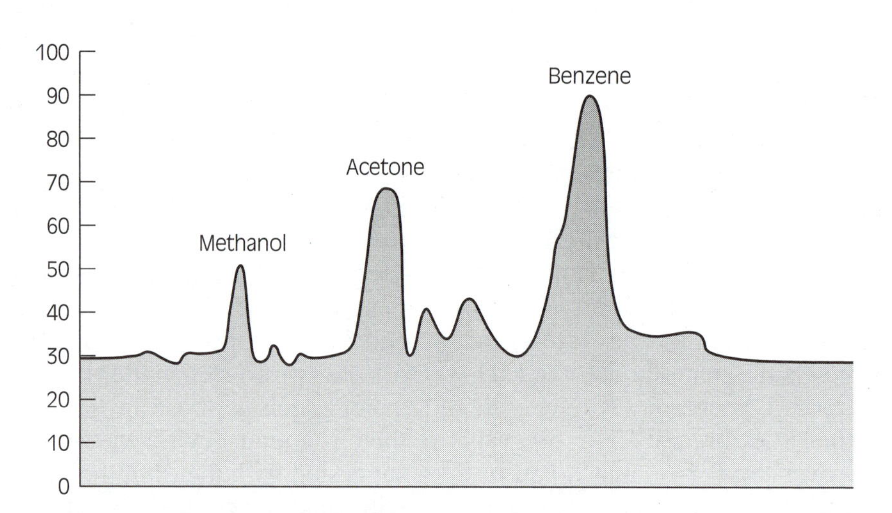

Figure 6-2 The GC/MS compares the spikes generated by the sample chemical with other chemicals spikes in its library, looking for a match.

ity although they are often set up to do drug screens. Other industrial locations usually have labs for quality assurance, and you should make arrangements to have samples run in an emergency. In most cases county or state environmental departments have labs that can run samples, but prior arrangements also need to be completed. When sending samples out, you need to determine what the lab prefers in the way of type of sample, sampling medium, and sample size. As litigation is always possible, it is best to split your samples between two labs and compare the results, and always be able to document the chain of custody.

SUMMARY

The use of ionizing detection devices is essential for responder safety, as they are valuable at measuring the potentially toxic portion of risk-based response. Most chemicals that HAZMAT teams respond to are flammable or combustible. Some of these chemicals are toxic (a few severely so) long before they reach their fire risk. The sensitive nature of the newer ppb models will provide valuable clues to possible chemical hazards during responses. Learning what types of materials can be measured, what levels can be expected, and the limitations of the detection devices are all crucial in the successful use of ionizing detectors.

KEY TERMS

FID See *flame ionization detector.*

Flame ionization detector A device that uses a hydrogen flame to ionize a gas sample; used for the detection of organic materials.

Gas chromatography A detection device that splits chemical compounds and identifies them by their retention and travel times.

GC See *Gas chromatograph.*

Ionization potential The ability of a chemical to be ionized, or have its electrical charges separated and measured. To be read by a PID, a chemical must have a lower IP than the lamp in the PID.

IP See *ionization potential.*

Mass spectrometer Almost always coupled with a GC, the MS measures the weight of a given unknown material and compares it to a library of known materials.

MS See *Mass spectrometer.*

Photoionization detector A detector that measures organic materials in air, by ionizing the gas with an ultraviolet lamp.

PEL See *Permissible exposure limit.*

Permissible exposure limit An occupational exposure limit established for an eight-hour period by OSHA.

PID See *photoionization detector.*

Threshold limit value An occupational exposure limit for an eight-hour day issued by the American Conference of Governmental Industrial Hygienists (ACGIH).

TLV See *threshold limit value.*

CHAPTER 7

COLORIMETRIC SAMPLING

- Introduction
- Colorimetric Science
- Detection of Known Materials
- Detection of Unidentified Materials
- Street Smart Tips for Colorimetric Tubes
- Additional Colorimetric Sampling Tips
- Summary
- Key Terms

INTRODUCTION

Colorimetric tubes are used for detecting known and unidentified gases or vapors. Colorimetric tubes are an easy method to narrow down an unidentified material to its chemical family. With an experienced user, specific identification can be made. It is unfortunate that this valuable tool in risk assessment is often overlooked or used incorrectly. **Colorimetric** sampling is not a simple task, and the user must be familiar with every aspect of the unit being used. The results one can obtain are invaluable. Even if one cannot determine what material is present, colorimetric tubes can tell us what is not present, which at the time can be quite important.

NOTE Colorimetric tubes are an easy method to narrow down an unidentified material to its chemical family.

NOTE Even if one cannot determine what material is present, colorimetric tubes can tell us which ones are not present, which at the time can be quite important.

COLORIMETRIC SCIENCE

Colorimetric sampling consists of placing a glass tube filled with a **reagent,** as shown in Figure 7-1, into a pumping mechanism, which causes air to pass through the reagent, usually a powder or crystals. The pumping mechanism can include bellow pumps, piston pumps, and thumb pumps. The type of pump used determines the pump stroke volume, stroke interval, and size of the tube. If the gas or vapor reacts with the reagent, a color change should occur, indicating a response to the gas sample. The three areas of change that are found are change of color, length of change, and color intensity. Detection tubes are made for a wide variety of materials and generally follow the chemical family lines (i.e., hydrocarbons, halogenated hydrocarbons, acid gases, amines). Although the tubes may be marked for a specific gas, they usually have cross-sensitivities (react to other materials), which at times is the most valuable aspect of colorimetric sampling. The tube systems may also have extension hoses, pyrolizers, prefillers, conditioning layers, sealed ampoules, pretubes and reagent tubes. It is very important that responders reading the instructions perform the sampling correctly with all the appropriate parts.

One colorimetric system that has been developed by Dräger involves the use of bar-coded sampling

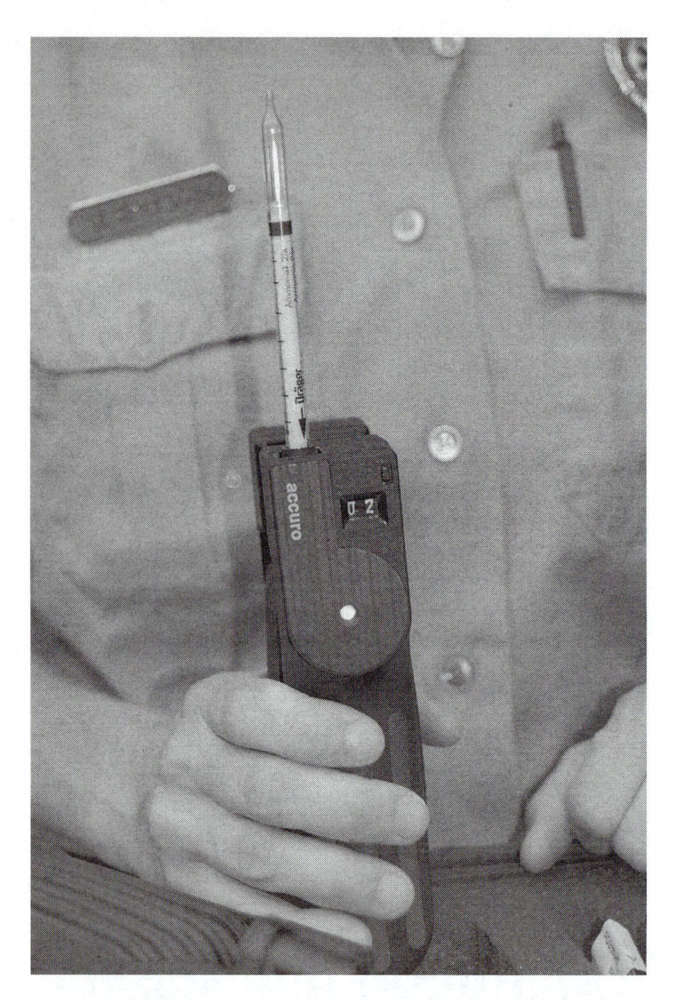

Figure 7-1 A typical colorimetric tube with a graduated scale in parts per million.

chips. This system, called the **chip measurement system (CMS)** and shown in Figure 7-2, involves the insertion of a sampling chip into a pump. The pump recognizes the chip in use and provides the correct amount of sample air through the reagent. By using light transfer through a reagent-filled ampoule, the pump provides an accurate reading of the gas that may be present. Currently one of the limiting factors in the use of the CMS is that there are not as many chips available as there are standard tubes. The advantage to this system over regular colorimetrics is that the reading is provided on a liquid crystal display (LCD) screen in parts per million, so it does not need interpretation. The other colorimetric systems require a length and color interpretation by a human, something that is at times faulty. The disadvantage to the CMS is that a general sensing tube is not available yet, nor can it sample more than one material at a time. Some standard colorimetric systems have the capability to increase the sensitivity of a particular tube or to reuse a tube after negative results, but the chip system does not allow this. The chips have ten sampling capillaries,

Figure 7-2 The chip measurement system (CMS) made by Dräger that takes a bar-coded chip and performs colorimetric sampling without human interpretation.

which after each is opened cannot be reused, nor can the sampling range be enhanced.

DETECTION OF KNOWN MATERIALS

If you know the material you are sampling for, there is a good possibility that a tube is available for that material. Most tubes are set up to read in parts per million, while others may be in percent by volume, and, even less common, some tubes just indicate presence. Most manufacturers provide a variety of sampling ranges for responders' use and a variety of sampling devices is available. When only one set of tubes is available it is possible to further enhance this range by increasing or decreasing the amount of sample air that passes through the tube. With all colorimetric sampling, make sure to read the instructions, and keep in mind that these tubes are calibrated for a certain temperature and humidity. The manufacturer provides temperature and humidity correction factors that should be used when using the tubes outside their normal operating parameters.

NOTE When looking for an unidentified, knowledge about colorimetric sampling, chemistry, and street sense is absolutely essential.

SAFETY Using colorimetrics we can come close to identifying the chemical family and its characteristics, and the potential hazards of that material. Even more important, we can determine whether it is immediately dangerous to life and health.

DETECTION OF UNIDENTIFIED MATERIALS

Detecting unidentified materials is not an easy task. When looking for an unidentified, knowledge about colorimetric sampling, chemistry, and street sense is absolutely essential. It takes more than one person to complete an investigation of an unidentified, one to sample, one to hold tubes, one to track results, and one to figure sampling strategy. Most manufacturers provide a flow chart that provides a guideline to follow when looking for unidentified materials. The flow charts guide responders to a chemical family and hopefully to narrow the choices of tubes and possible chemicals present. There is not a monitor that exists that will absolutely 100 percent indicate that "x, y, z" is the material present. However, using colorimetrics we can come close. We can identify the chemical family, its characteristics, and the potential hazards of that material. Even more important, we can determine whether it is immediately dangerous to life and health. One of the best things that a colorimetric system does is rule out what is not present. If you go through an entire standard colorimetric tube set, you have sampled for a wide variety of chemical hazards and have effectively covered the whole spectrum. Table 7-1 provides an overview of an unidentified flow chart from two manufacturers. The Dräger system requires several tubes to work through the various families. The Sensidyne system has four **polytech** tubes that are used to provide "hits" for possible substances. The polytech I tube is to be used much like the Dräger **polytest** tube and to then follow their unidentified flow chart, which somewhat resembles the Dräger chart but has some variations and different tube names. A street smart tip: Use the Sensidyne polytech I and polytech IV to start your search, then follow the search pattern provided in Chapter 10 for unidentified materials.

The one major key point when using colorimetric tubes is to be familiar with the instructions that come with each tube. No one can memorize the instructions for each tube and its varying requirements and problems. Responders should know how to obtain this information from the instructions with each tube. It is recommended that the department make a book of instruction sheets to make searching easier. The instruction sheet lists the number of pumps or strokes for the tube, the color change to be anticipated, and, most importantly, the cross-sensitivities and associated problems with the tube. When sampling for unidentified materials, cross-sensitivity is most important as it may be your only clue as to the material present. The tubes are manufactured with filters, screens, and other materials at each end. A material for which the tube is sensitive may only change the filter and not change any other part if the tube. If you miss this change, it may be the only "hit" you get for that material. If you get a change on a tube that is not indicated by the instructions (either in the indicating layer information or the cross-sensitivities listing), one can only assume that there is something there but identification by this tube is not possible. Only use color changes that are indicated somewhere on the instruction sheet.

NOTE When sampling for unidentified materials, cross-sensitivity is most important as this may be your only clue as to the material present.

STREET SMART TIPS FOR COLORIMETRIC TUBES

The instructions with colorimetric tubes and some other texts mention that the tubes can be inaccurate up to 25 percent of the time, sometimes even higher. This inaccuracy rate varies tube to tube and even the instructions provide a varied rate. During testing some tubes that are years out of date and have been stored in unusual conditions have been found to be up to 98 percent accurate. But for liability reasons it is best to follow the manufacturer's provided variance factor. Many factors can cause this inaccuracy, some of which are provided in the following list:

- Humidity plays a major factor in the response of some tubes and most tubes are calibrated to 50 percent humidity. If the humidity is different, then you must factor in the increased or decreased humidity.
- Atmospheric pressure also plays a role, as the tubes are calibrated to a certain pressure. If the atmospheric pressure is different from what

TABLE 7-1

Unidentified Flow Chart for Colorimetric Tubes

Dräger Unidentified Flow Chart

Tube	Chemical Family	Reaction/Notes
Polytest	Organics and inorganics	Positive means that something is present; the higher the reading the more material that is present. A negative reading does not mean the air is safe.
Halogenated hydrocarbon	Halogenated hydrocarbons	Most common would be freon, but others include methyl chloride, dichloromethane, trichloromethane, chloroform, carbon tetrachloride
Ethyl acetate	Organic substances	Rules out inorganics
Benzene	Aromatic hydrocarbons	Benzene, toluene, ethyl benzene, xylene, cumene, and styrene
Acetone	Ketones	Acetone, methyl ethyl ketone, methyl isopropyl ketone, and methyl isobutyl ketone
Alcohol	Alcohols	Include isopropyl alcohol, ethyl alcohol, methyl alcohol, ethylene glycol, propylene glycol, propyl alcohol, 1-propanol, 2-propanol, allyl alcohol, 1-butanol, 2 butanol, tert-butyl alcohol
Ammonia	Amines and ammonia	Amines, ammonia, and hydrazine
Formic acid	Organic acid gases	Color change indicates type of acid that may be present

Sensidyne Gastech Colorimetric Tubes

Tube	Chemical Substances	Reaction/Notes
Polytech I	Carbon disulfide Hydrogen sulfide Carbon monoxide Acetone Acetylene Ethylene Benzene Propane, propylene Styrene Trichloroethylene Gasoline Toluene, xylene	No differentiation is possible with some of these combinations, although among the major groups there are color differences. As an example both carbon disulfide and hydrogen sulfide cause a green color change.
Polytech II	Ammonia, amines Sulfur dioxide Hydrogen sulfide Carbon monoxide	Also does hydrogen chloride, chlorine, nitrogen dioxide, and phosphine.
Polytech III	Ammonia Hydrogen sulfide Hydrocarbons	Also does sulfur dioxide, hydrogen chloride, chlorine, nitrogen dioxide, butane, gasoline, and liquid petroleum gas (LPG).
Polytech IV	Ammonia, amines Hydrogen chloride Hydrogen sulfide Chlorine Carbon monoxide Carbon dioxide	Also does sulfur dioxide, nitrogen dioxide, acetylene, ethylene, phosphine, hydrogen, methyl mercaptan, and propylene

the tube is calibrated for, the instruction sheet provides a formula to determine the actual reading.

- Each tube has a specific operating temperature range; most are good for 32°F to 122°F.
- Age, light, and storage conditions also determine the accuracy of the tubes. All tubes have a specific shelf life, and because the tubes work on a chemical reaction and corresponding color change principle, light can cause a change within the tube. Although some manufacturers recommend that the tubes be stored in the refrigerator, if you will be taking them in and out frequently, it is best to store them in conditions that do not have temperature extremes.
- More than one gas present in a sample area can alter the readings. Remember that the tubes react to many other gases than just what the box indicates. It is important to know the cross-sensitivities of each tube you use.
- It is important to know what the color of the tube is prior to using it in the pump so you can recognize a color change. Having the sampling team keep an unused, closed tube for comparison is a good idea.
- Smoke from various tubes is usually water vapor and is harmless, but in some cases other gases may be released. It is not uncommon for tubes to heat up and smoke, depending on the gas being sampled. One particular chemical tube creates an extremely hot chemical reaction, which in a flammable atmosphere is a perfect ignition source. When using colorimetric tubes it is important to check for the presence of flammable gases prior to using some colorimetric tubes. The tubes that could cause problems are rare, but one common tube that heats is the Dräger halogenated hydrocarbons tube. With some off-gases, persons with respiratory problems could be affected.
- It is important to clear the pump between tubes to clear the previous sample. As chemical reactions take place within the tubes, it is essential to ensure a clean pump for the next sample.

ADDITIONAL COLORIMETRIC SAMPLING TIPS

By increasing or decreasing the amount of air passing through the tube, you can increase or decrease its sensitivity and some examples are provided in the case study Ammonia and Tube Colors. By changing this sensitivity you do not need to maintain several boxes of tubes. Even when you do not get any change on a tube, by changing its sensitivity, you may get a "hit" for that gas. As an example, the ammonia tube takes ten pumps and measures 2

CASE STUDY

Ammonia and Tube Colors—Most of my unusual incidents with colorimetric tubes have involved ammonia. In one incident we responded to a reported ammonia leak at a chicken processing plant. I was setting up the tubes in preparation to go into the building when I snapped the ends of the ammonia tube, and it instantly changed colors. We quickly changed the isolation distance another block or two. In another case we had gone to assist an engine company with a cylinder buried under an apartment complex. The woman who lived in the basement apartment had been smelling ammonia, and the rental company had been trying to locate the source. They had torn the woman's apartment apart, including the walls, floor, and ceiling. When they did not find anything in the apartment, they decided to dig near the foundation where they found the cylinder buried under the foundation. The apartment building had been there at least twenty-one years, so the cylinder had been there at least that long. The cylinder looked like a nurse ammonia tank a farmer would have used. The apartments were built on an old farm, so it is suspected that the cylinder was left behind from the farm and was buried with the other backfill. We grabbed a colorimetric tube for ammonia and proceeded to the hole, with my partner jumping in. He started pumping the colorimetric tube, and very quickly I asked him, when did it change? His reply was it had not changed. I was not sure what color it was to begin with, so I grabbed a new tube to compare. It had changed colors, but we did not know when, so I prepared another tube and handed it to my partner, and on the first pump stroke it went off scale. With all the information we were pretty sure the cylinder was ammonia. We dug the cylinder out, used the bomb technicians' trailer to move the cylinder to a remote location where we opened the cylinder in a porta-tank of 2500 gallons of water, and made a large amount of ammonium hydroxide solution.

ppm to 30 ppm. If after pumping ten times you did not get any reading, the amount of pumps or strokes can be increased by another ten and that will increase the sensitivity to a range of 1 ppm to 15 ppm. When sampling in this fashion, any readings are estimated but are valuable when searching for an unidentified. Many times small amounts of chemicals in the air present a chronic hazard and are irritating over an extended period of time. However, if pumping the ammonia tube ten times results in a reading that is off the scale, then we can decrease its sensitivity by limiting the number of pump strokes. If we pumped the tube five times, and the tube is reading 4 ppm to 60 ppm, pumping the tube one time would allow a reading of 20 ppm to 300 ppm. In some cases the color change is very quick, so it is important to know what the original color of the tube was. For some chemicals the tube will change dramatically without any air being pumped through the tube and will provide a quick response to those gases. The instructions must be read to ensure that there are no special restrictions on the tube used in this fashion. Some of the chemical warfare agent tubes have inner ampoules that are broken after the air is passed through the tube and could not be used in this fashion.

It is important to know some basic chemistry, as this is invaluable when dealing with unidentified materials in the identification flow charts. In the event you did not get any hits on the flow chart, your next option is to go through every tube in the kit, watching for any possible hits. In Chapter 10 there are some example sampling strategies using colorimetric tubes. It may be best to combine tubes from the various manufacturers in your system. By combining several tubes, you can shorten the time required to do some broad range sampling. Although there has been considerable discussion in the response community about using one manufacturer's tube in another manufacturer's pump. The best rule is that a fat tube goes with a pump intended for fat tubes and a skinny tube is used in a pump for skinny tubes. Most tubes require 100 cc's of air to be passed across the sampling media, and the tube does not care if that is pulled across through a piston pump or a bellows pump. When trying to identify an unidentified gas in a characterization mode, the pump type is a moot issue. When doing occupational exposure, evidentiary, or other legal sampling, then it is advisable to use the manufacturer's recommended pump. Because the sensitivity varies with colorimetrics, a slight change is a possible clue. Also a color change in a prefilter is a hit and may indicate a possible gas. When looking for an unidentified gas, many times we can find unusual sources, so check for all possible gases available. When you have gone through the whole sampling process and have found nothing, what you can say is: this, this, and this were not present at the time of sampling.

SUMMARY

Although colorimetrics are one of the more confusing detection devices, they can provide the greatest amount of information. They can take an unidentified situation and bring it to a chemical family, if not to a specific identification. They are time consuming and require the user to read the instructions. No one colorimetric system is better than the other, and in fact the best system is a combination of several manufacturer's systems.

KEY TERMS

Chip measurement system A colorimetric sampling system that uses a bar-coded chip to measure known gases.

CMS See *chip measurement system.*

Colorimetric A form of detection that involves a color change when the detection device is exposed to a sample chemical.

Colorimetric tubes Glass tubes filled with a material that changes color when the intended gas passes through the material.

Polytech The name for sensidyne's unknown gas detector tube.

Polytest The name for the Dräger unknown gas detector tube.

Reagent A chemical material (solid or liquid) that is changed when exposed to a chemical substance.

RADIATION DETECTION

INTRODUCTION

Dealing with radiation incidents makes many responders uncomfortable. Emergency responders do not have an in-depth understanding of radiation and its applicable hazards. With terrorism on the rise, responders need to be comfortable with the detection and monitoring for radiation. Although **nuclear detonation** is very unlikely, the use of a **radiological dispersion device (RDD),** in which a conventional explosive is used to distribute radioactive materials, is not beyond imagination. Any number of potential sources of radiation could easily be used for this purpose. Responders need to be comfortable with their skills in handling radiation events, so that the community does not panic.

SAFETY With terrorism on the rise, responders need to be comfortable with the detection and monitoring for radiation.

CASE STUDY

In a small community, near the end of the school year, a high school demonstration of a radiation monitor resulted in an interesting response to a radioactive rock. The teacher was checking the radiation levels of various items, and the monitor started to click loudly when placed near a large rock that had been in the classroom for many years. The HAZMAT team was called, and wearing lead aprons, placed the suspect rock into a lead pig. The school was closed for the remainder of that day and the next, and the children were advised to take showers and wash their clothes and shoes. The rock, which was from an unknown location, contained uranium, a common material. The issue at hand is that the detector used was very sensitive and the rock did not present any risk to the students. Most smoke detectors (americium) or exit signs (tritium) emit more radiation. The responders did not understand the readings, nor were they able to correlate them to what was safe or dangerous.

In another incident, a perceived threat was provided to several federal agencies stating that they would receive radioactive packages. In the next couple of days, several packages arrived in a number of different locations. In one case, responders misread the readings on their monitor and indicated that the package was providing readings into the millirem range, which would cause some concern as that would be unusual. The reading was actually microrem and was not much higher than the background level.

Radiation detectors for emergency responders are divided into two major groups: one measures exposure to radiation and the other measures the current amount of radiation in the area. To effectively measure radiation at least two detectors are needed as no detector measures all three types of radiation.

NOTE Radiation detectors for emergency responders are divided into two major groups, one measures exposure to radiation and the other measures the current amount of radiation in the area.

TYPES OF RADIATION

Radiation is comprised of two basic categories: ionizing radiation and nonionizing radiation. Some examples of nonionizing radiation include radio waves, microwaves, infrared, visible light, and ultraviolet light. Alpha, beta, gamma, and X rays are forms of ionizing radiation, and some characteristics of each are provided in Figure 8-1. There are two subcategories of ionizing radiation, one with energy and weight and the other comprising just energy. Alpha and beta are known as particulates and have weight and energy. Gamma radiation has just energy and no weight.

NOTE Alpha, beta, gamma, and X rays are forms of ionizing radiation.

Alpha is a particulate and only has the ability to move a few feet. Its primary hazard is through inhalation or ingestion. When monitoring for alpha, responders need to monitor very slowly as alpha does not have much energy. Scanning a person's body for alpha should take several minutes. Street clothing and other PPE provides ample protection against alpha radiation.

Beta radiation is of two forms, low and high energy, but both are particulates. The low energy form is comparable to alpha, but has the ability to move a little farther and cause more damage. High energy beta moves even farther and can cause greater harm. Beta can move several feet and higher levels of PPE are required, but turnout gear offers some protection.

Gamma is not a particulate, but is airborne energy described as wavelike radiation that comes from within the source. These waves are sometimes called *electromagnetic waves* or *electromagnetic radiation*. Gamma can move a considerable distance. Lead offers some protection. It is high energy and can cause internal damage to the body without causing external damage.

X rays are comparable to gamma radiation waves, as they are wavelike, but X rays are only dangerous when the X-ray machine is energized. The radiation from an X ray is caused when speeding electrons strike other electrons in a target. When they collide the amount of energy created emits powerful waves of radiation. Under normal circumstances this collision occurs in an X-ray tube and is focused on a target, usually a photographic plate of film. When the energy hits an object between the source and the photo film, it interacts with it. If it strikes an electron it rips the electron from the atom, the wave then has weight and moves on to the target. The more dense the object, the more waves are absorbed. For example, bone is less dense material so it allows more X rays to reach the film resulting in a negative image. Therefore, on the film the bone is light, and the flesh is dark. This ripping causes a chemical reaction, which is why X rays and gamma radiation are much more damaging than alpha or beta.

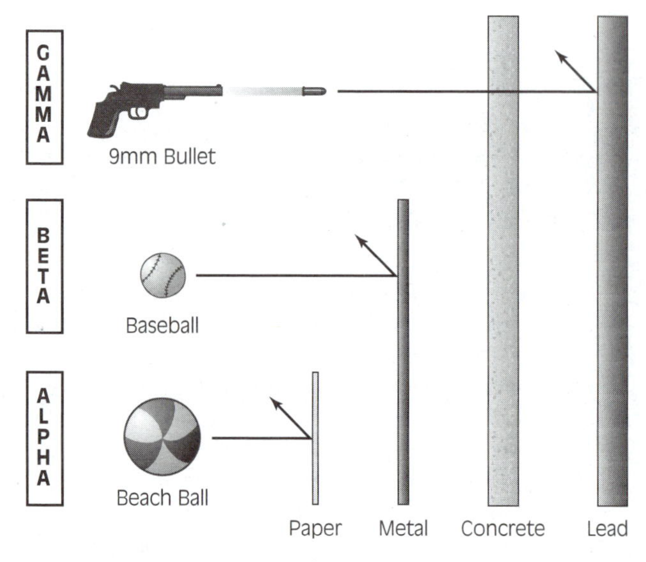

Figure 8-1 Examples of the various strengths of radioactive energy sources.

RADIATION PROTECTION

The basis for protection from radiation is time, distance, and shielding, a concept that should be applied to all types of chemical exposures as well. We need to understand this concept as the meters generally provide measurements based on a timed exposure. Exposure levels are based on time as well. Time is important as in many cases, humans can sustain a short exposure to a radiation source without being harmed. If you want to limit your radiation exposure to 1 millirem per hour (mR/hr) and the source is reading 60 mR, you could be near the source for one minute and your exposure would be 1 mR/hr. Being near the source for an hour results in an exposure of 60 mR/hr. Distance from the source also plays a factor, and the safety factor with distance is provided in the fact that exposure levels are based on an inverse square law. If the source has a radiation level of 20 mR/hr at 2 feet then moving back to 4 feet provides and exposure of 5 mR/hr. Moving to 6 feet would result in an exposure of 1.25 mR/hr.

SAFETY The basis for protection from radiation is time, distance, and shielding, a concept that should be applied to all types of chemical exposures as well. We need to understand this concept as the meters generally provide measurements based on a timed exposure.

There are several varied methods of measuring radiation that are commonly used. There has been an attempt to change the measurement systems for several years, but this change has not been made throughout the industry. Table 8-1 provides a conversion for the various units. The basic unit of measuring radiation is the **roentgen** (R or r), which is a value provided for the amount of ionization in air caused by X rays or gamma radiation. The most common method of measuring a radiation exposure or dose is in a term known as **REM (R),** which stands for roentgen equivalent in man. The values 1 roentgen and 1 REM are used interchangeably and are equivalent. The radiation dose is important, but equally important is how quickly the dose is being applied. The measurement of this speed is the dose rate, usually expressed in an hour. The formula is Dose = Dose rate × time. As an example, suppose you are at an event where the radiation meter is reading 1 R/hr, how long would it take to receive a dose of 400 mrem?

1 R/Hr = 1000 mR/Hr
400 mrem = 1000 mR/hr × time

Dose = Dose rate × time
time = 400/1000 = 0.4

0.4 × 60 min/1 hr = 24 mins

It would take twenty-four minutes to receive a dose of 400 mrem.

A variety of terms is used to describe radiation levels, many of which are provided here. A numerical value in REM would be a high level of radiation, and most radiation sources we deal with are in the **millirem** scale (mR). Normal background radiation is in **microrem** (μ R). The differences between the three are mathematical, as 1 REM is 1000 times more powerful than a millirem, and 1,000,000 more powerful than a microrem. Another term that may

TABLE 8-1

Radiation Equivalents and Conversions

Term or Amount	Equal To
RAD	Gray (Gy) or Absorbed dose (AD)
Rem	Sievert (Sv) or Dose Equivalent (DE)
1 μ rem	0.01 μ Sv
100 μ rem	1 μ Sv
1 mrem	10 μ Sv
100 mrem	1 mSv
1 Rem	10 mSv
100 Rem	1Sv
Curie (Ci)	Becquerel (Bq)
27 picocuries	1 disintegration/sec (d/sec) or 1 Becquerel
1 picocurie	37 milliBq
1 μCi	37 killoBq
27 μCi	1 MegaBq
1 mCi	37 MegaBq
27 mCi	1 GigaBq
1 Ci	37 GigaBq
27 Ci	1 TeraBq
1 pCi	2.22 dpm
1 Bq	60 dpm
1 dpm	0.45 pCi

be used is **RAD,** which is radiation absorbed dose, and is a quantity of radiation. The term used to describe the amount of radiation is described as the source's activity, which is the amount of time for the material to decay. The measurement for this activity is in **curies** which is the base unit, but there are also millicuries and microcuries. It is not uncommon to find these values on shipping papers, because when shipping radioactive materials there can be anticipated radioactivity coming from a package. The new International System (SI) method of measuring radiation uses **Gray** and **Sievert** as the common terms. Their equivalents are provided in Table 8-1.

Now that we have determined the various types of radiation and their measurement terms, we need to establish some action criteria. As was mentioned previously time, distance, and shielding must be taken into account. The action levels provided in Table 8-2 are assuming that the person is not wearing any protection and is a whole body exposure unless noted. There are three types of action levels as well, one is for the general population, one for emergency responders, and one for nuclear industry workers. There is always a risk when dealing with any chemical substance. Think of these action levels much like you think of PELs and TLVs.

RADIATION MONITORS

Current radiation monitors provide two methods of measuring radiation: REM and count per minute (CPM), with the REM being the most important to emergency responders. A variety of detection devices is available, but the important consideration is the probe that is attached to the unit. The probe actually determines what type of radiation can be detected by the detector. The monitor shown in Figure 8-2 has a probe that detects alpha and beta radiation, and an internal probe that measures gamma radiation. A variety of probes is available, but the most common is a pancake probe that is useful for small amounts of radiation. When the amount of radiation becomes higher, the responder then switches to the internal probe, which is designed for higher levels of radiation. When turned on, a radiation monitor will pick up naturally occurring background radiation, which should be in microrem. The amount of background radiation varies from city to city, and

TABLE 8-2

Radiation Action Levels and Assorted Doses

Dose	Cause or Type	Note
	Average exposures to radiation	
0.01 rem annually	Chest X ray	Per year exposure
0.2 rem annually	Radon in the home	Per year exposure
0.081 rem annually	Living at high elevations (Denver)	Per year exposure
1.4 rem	Gastrointestinal series	
	Action Levels	
1 mrem/hour	Isolation zone (public protection level)	Is recommended exposure limit for normal activities
5 REM	Emergency response	For all activities
10 REM	Emergency response	Protecting valuable property
25 REM	Emergency response	Lifesaving or protection of large populations
>25 REM	Emergency response	Lifesaving or protection of large populations. Only on a voluntary basis for persons who are aware of the risks involved.

Source: National Fire Academy Emergency Response to Terrorism Tactical Considerations HazMat Student manual, p. SM 3-25 February 2000.

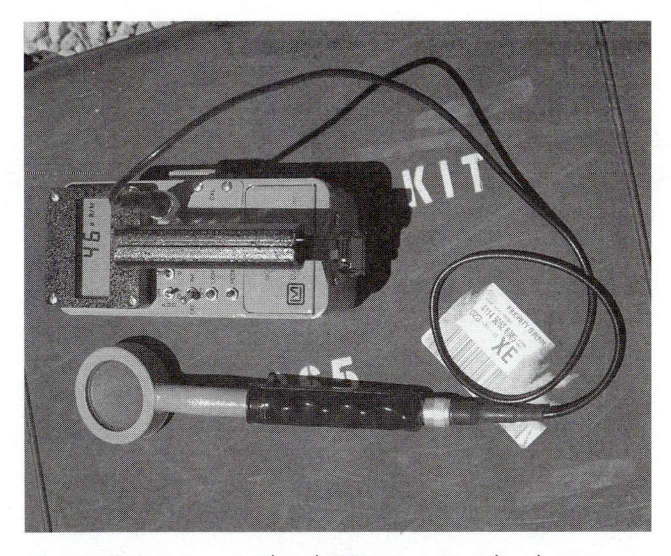

Figure 8-2 A typical radiation monitor that has a pancake probe for alpha and beta. The gamma probe is inside the instrument.

even varies within a few miles. It is important to do some testing to determine the normal levels in various parts of your community. Then responders can know when they are being exposed to radiation levels higher than background.

The technology for this detection is based on the radioactive particles, which have a particular electrical charge associated with them entering the sensing chamber. The most common detection technology is known as **Geiger-Muller (GM)** and an ionization chamber. The Geiger-Muller is used for low energy materials and the **ionization chamber** is used for higher levels.

Another method of radiation detection is the use of a **radiation pager** (Figure 8-3) or a **dosimeter**

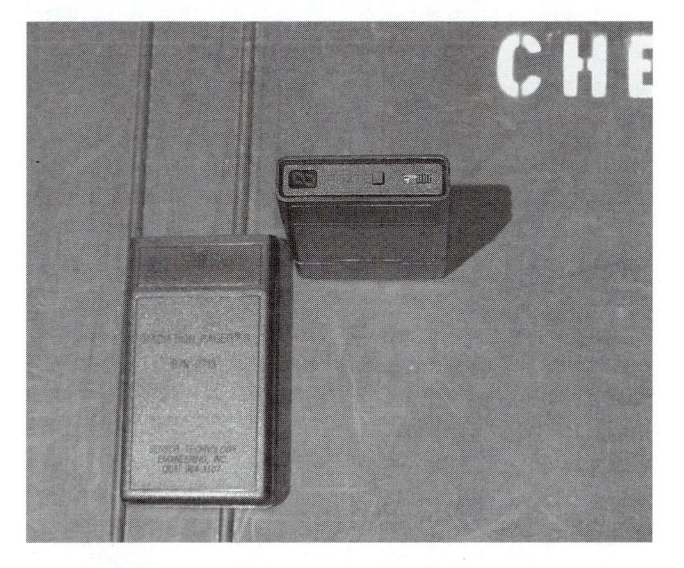

Figure 8-3 A radiation pager that alerts when a level above background is reached.

Figure 8-4 An electronic radiation dosimeter, that measures the dose of radiation a person receives. This unit also downloads the dose information to a central location for personnel tracking.

(Figure 8-4). They are two separate devices, but are commonly used for the same purpose. The radiation pager is used to ensure the safety of personnel operating in environments where radiation exposure is possible. Radiation pagers detect X rays and gamma, but they also detect high-energy beta radiation. When the pagers are turned on, they calibrate themselves to the background radiation. If a responder enters an area that has radiation levels above background, the pager will alert, either through an audible alarm or vibration. It provides a reading of one to ten times above the background level, which can be correlated to a range of radiation, based on the initial background radiation. It is very important especially when dealing with terrorism or criminal events that you check the level of background radiation to make sure you are in a safe area when you turn on the pager. Someone intent on harming responders could place the radioactive source at the area where crews would arrive and set up. When they turned on the pagers they would set themselves up as the pagers would record a high level as normal background. The other form of protection is a dosimeter, which records the dose that your body is taking. Older dosimeters just recorded the dose, and the user had to visually check the dosimeter to see the exposure. New dosimeters are electronic and exist as pagers. They not only record the dose, but also measure the amount of radiation that exists. They also alarm when the user enters certain levels of radiation. These dosimeters provide protection for X ray, gamma, and beta radiation.

NOTE These devices are sensitive enough to detect radiological pharmaceuticals such as those used in chemotherapy, so it may alert if in close proximity to someone who recently had chemotherapy.

Through the years, radiation has been forgotten and has not been a priority for sampling. With HAZMAT teams facing terrorism concerns and high tech criminal events, there is a need to check for radiation hazards more frequently. According to news reports and FBI records, radiation events are on the increase. According to the Nuclear Regulatory Commission many people are murdered using radioactive sources each year, and a number of radioactive sources are stolen each year. The most common theft is of ground imaging devices, which in some cases have significant radioactive sources that can cause harm to a community. The records of the NRC indicate that there are about 9000 devices missing in this country. When dealing with unidentified materials, packages, containers, drum dumps, and other possible criminal activities, it is very important to check for radiation. The simplest form of protection is to use a radiation monitor, establish a background level, then turn on the radiation pager. Crews should get accustomed to using these devices on a regular basis to become comfortable with them and to learn where radioactive substances exist. One warning though is that these devices are sensitive enough to detect radiological pharmaceuticals, such as those used in chemotherapy, so it may alert if in close proximity to someone who recently had chemotherapy.

SAFETY When dealing with terrorism or criminal events, it is very important that you check the level of background radiation to make sure you are in a safe place when you turn on the pager.

SUMMARY

HAZMAT responders need to become familiar with radiation detection, as the possibility exists for future events. Many radioactive substances exist, and anytime responders are dealing with unidentified materials, they should check for radiation. Knowing how to monitor for radiation is as important as knowing what the action levels and measuring ranges are for the various types.

KEY TERMS

Alpha Type of radioactive particle.

Beta Type of radioactive particle.

Curie Measurement of radioactive activity level.

Dosimeter Device that measure the body's dose of radiation.

Gamma Form of radioactive energy.

Geiger-Muller Detection device type for low energy materials.

Gray Measurement of radioactivity, equivalent to RAD.

Ionizing radiation Radiation that has enough energy to break up chemical bonds and can create ions. Examples include X rays, gamma radiation, and beta particles.

Ionization chamber Detection type-device for high energy materials.

Microrem Measurement of radiation: normal background radiation is usually in microrems.

Millirem Higher amount (1000 times) of radiation than microrem.

Nonionizing radiation Radiation that does not have enough energy to create charged particles, such as radio waves, microwaves, infrared light, visible light, and ultraviolet light.

Nuclear detonation Device that detonates through nuclear fission; the explosive power is derived from a nuclear source.

RAD Radiation absorbed dose, a quantity of radiation.

Radiation pager Detection device that alerts in the presence of gamma and X rays

Radiological dispersion device Explosive device that spreads a radioactive material. The explosive power is derived from a nonnuclear source, such as a pipe bomb, which is attached to a radioactive substance.

RDD See *Radiological dispersion device.*

REM Radiation equivalent in man, a method of measuring radiation dose.

Roentgen Basic unit of measurement for radiation.

Sievert Unit of measurement equivalent to REM.

X ray A form of radiation much like light but that bombards a target with electrons, which makes it very penetrating.

OTHER DETECTION DEVICES

- Introduction
- Mercury Detector
- HazCat Kit
- Test Strips
- Water-Finding Test Paper
- Drum Sampling
- Infrared Thermometers
- Summary
- Key Terms

INTRODUCTION

This chapter discusses a variety of detection devices, many of which have very specialized uses. Once a substance is characterized into a fire, corrosive, or toxic risk, further identification needs to be made. The use of these other detection devices are integral to this identification. The HazCat™ kit assists in the identification of unidentified materials and can be of great assistance at a drum dump or other environmental crime. Not all HAZMAT teams may have the budget available to purchase some of these more expensive items, but it is important to understand what kind of information these items can provide and to know where you can obtain the equipment if it were needed. A mercury detector is a great example. Not every HAZMAT team requires one, but knowing what it can do for you and where to get one is key to ensuring a safe response to a mercury release.

Figure 9-1 A mercury detector.

MERCURY DETECTOR

A mercury detector (Figure 9-1) is used to detect mercury vapors in the air. One unit has a measuring range of 0.001 to 0.999 mg/m^3 of mercury, and Chapter 1 discusses mg/m^3 conversion to parts per million. The unit has a sensor lined with gold foil, which attracts mercury (Figure 9-2). If the unit is placed in an environment that has mercury vapor in the air, the mercury will be attracted to the gold foil. As the mercury collects on the gold foil, the electrical resistance changes and causes a corresponding reading on the display. This type of detector is very susceptible to false readings if the instrument is hand carried or is moving around. The best situation is to have the instrument placed on a table or other stable environment and bring the potential contaminant up to the instrument. Another method is to take

Figure 9-2 Using a magnifying glass looking for mercury. A flashlight shined at a right angle assists in this process.

three readings and average the readings to arrive at a value that should be close to the actual level. When dealing with possible victim contamination, it is best to remove their clothes, bag them, and then later insert the probe of the instrument into the bag to sample the atmosphere. HAZMAT responders do not encounter mercury on a regular basis, but since it is a chronic health hazard and a very toxic material through chronic skin contact, a detection method should be available, and you should have an emergency contact to obtain this type of device. A low cost alternative is colorimetric tubes that can detect mercury vapors. These tubes have a detection range of 0.05–2 mg/m^3 for Dräger and 0.05–2.5 mg/m^3 for Sensidyne, which is more than adequate for sampling purposes. In many cases the material is spilled in a home, and a concern for the health and welfare of the residents needs to be a high priority. Some terrorists use mercury in explosive devices, so mercury may be present at a terrorism incident.

CASE STUDY

A 911 call was received for a mercury spill, in which a barometer had been broken, and the mercury had spilled out. Upon arrival at a local retirement community, a moving crew was noted and the people who had called 911 were located. It was determined that a large barometer had been broken during the move, and that when it was unpacked mercury had spilled out in the resident's apartment (Figure 9-3). The apartment was the last apartment on the other side of the building, a considerable distance away from the front of the truck. It was possible that mercury was spilled throughout the hallways leading out of the building, and since this pathway led to other areas of the building, the mercury could be throughout the building. The release of mercury even in small quantities is a big concern; it is not an acute health hazard but is a serious chronic health hazard. The Agency for Toxic Substances and Diseases Registration (ATSDR) has a web site located at http://www.atsdr.cdc.gov/toxprofiles/phs8916.html that has considerable information. There is not a need for extensive PPE; just avoid contact. The inhalation of vapors is a long-term issue, not a short-term problem. Isolation of the contaminated areas is essential to make sure the mercury is not tracked elsewhere. The persons who were in or near the spill were isolated, and those who it was suspected got the mercury on them were told to remove their clothes, and were provided other clothes. The clothes were bagged and tagged and placed outside into the sun (Figure 9-4). The Maryland Department of Environment (MDE) has the only mercury detector in the state, so it was requested for assistance (Figure 9-5). Complicating this was the fact that about ten people had been in and out of the apartment looking at the spill. One person who likely had mercury on him, had driven to a local store to get some food, which had the potential for moving the problem off site. By the time the monitor arrived, we had established a monitoring plan to check for contamination. We checked the hallways and luckily did not find any contamination. We checked the apartment where there was a visible mercury spill and found this to be the only location where there was any mercury. We then checked and found four sets of clothes that had been contaminated, and four sets of shoes that were contaminated; the remaining clothes were clear. During this process we found a worker from another building who had been in the room and had cleaned up some of the mercury and placed it in a bio waste container for disposal. Some of the tools used to clean up the spill were put in one of several dumpsters. It was also determined that the box that held the barometer had been taken back outside and placed with the other trashed boxes and packing material. All of this was bagged as waste for disposal, as it could not be determined what packing material was in the contaminated box. Unfortunately when mercury is spilled on carpet, the carpet must be picked up and destroyed, as there is no method of decontamination that would not degrade or destroy the carpet. The barometer was an antique barometer built in 1790 and was worth \$30,000. The carpet, also an antique, was worth \$100,000.

Figure 9-3 Visually looking for mercury while using the mercury detector.

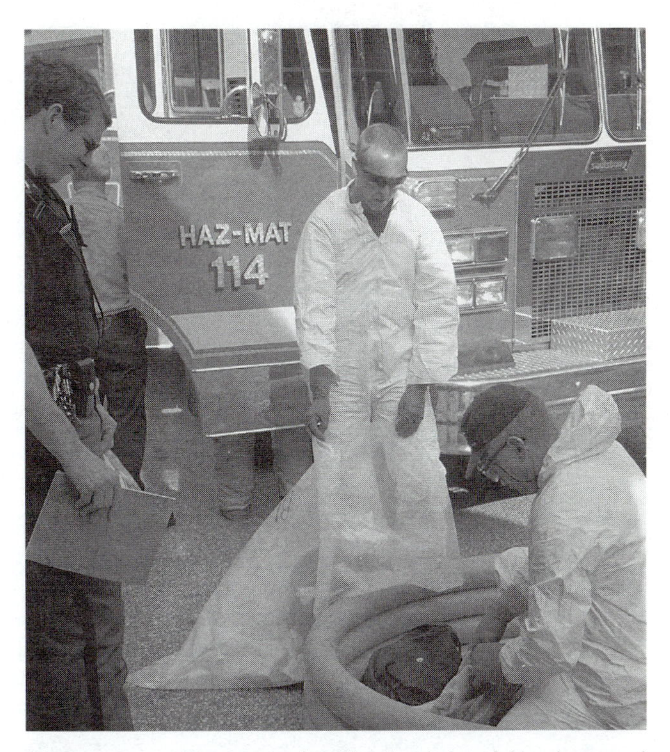

Figure 9-4 Clothes to be checked should be bagged and placed in the sun. When opened the bag should be in a containment area.

Figure 9-5 A police officer tracks the person's valuables while Bob Swann of MDE checks for contamination with the mercury detector.

Figure 9-6 Checking for contamination, as mercury has an affinity for gold. The mercury detector has a gold foil detection source.

 HAZCAT KIT

The **HazCat™ kit** is a system that uses a chemistry set to assist in the identification of unidentified solids, liquids, and gases. The HazCat kit uses colorimetric tubes to identify unidentified vapors. Information on the use of colorimetric tubes is provided in Chapter 7, so it is not repeated here. For many of the solids and liquids, the kit uses a flow chart to sort through the various chemical families, which would be the minimum identification. The HazCat kit does a great job of providing more specific identification. Much like the colorimetric system, the HazCat kit relies on the fact that chemicals within the same family react in the same fashion. By using a series of chemical tests, including mixing the unidentified with a number of chemicals called reagents. This system uses the process of elimination to identify and/or characterize a sample. The system requires specific training, and a background in chemistry helps. Frequent use or training is required to maintain proficiency. The system does very well to identify unknown materials, and responders who regularly use the kit, can make identification in about twenty to thirty minutes, while users who are unfamiliar with the kit take much

TABLE 9-1

Concerns with Unidentified Solid Materials
Water reactivity
Air reactivity
Explosive characteristics (oxidizers)
Peroxidized materials
Toxicity (cyanide, sulfides)
Sugar test
Flour test

longer. In emergency response, identify the risk the material presents, make the situation safer, and then under less stressful conditions, make the identification using the HazCat kit.

Typically the one big area that is deficient in effective sampling is the identification of unidentified solids. Some of the initial concerns with solids are found in Table 9-1 and some concerns with liquids are found in Table 9-2. Sugar and flour may seem unusual to you, but these two items are used fairly often to mimic terrorism agents in terrorist hoaxes, so these tests are included.

Provided in Table 9-3 are a few of the tests that can be accomplished by the HazCat kit for solid materials. The HazCat system has full flow charts and provides extensive documentation and a process for determining the identity of unidentified solids. These tests in the table provide a quick overview of some of the initial characterizations listed in Table 9-1 and are in no way intended to replace the HazCat system. Refer to the HazCat kit for additional safety concerns and more information.

SAFETY These tests are provided for your review as a tiny portion of the HazCat system, and are in no way intended to replace the HazCat system. Refer to the HazCat kit for additional safety concerns and more information.

TABLE 9-2

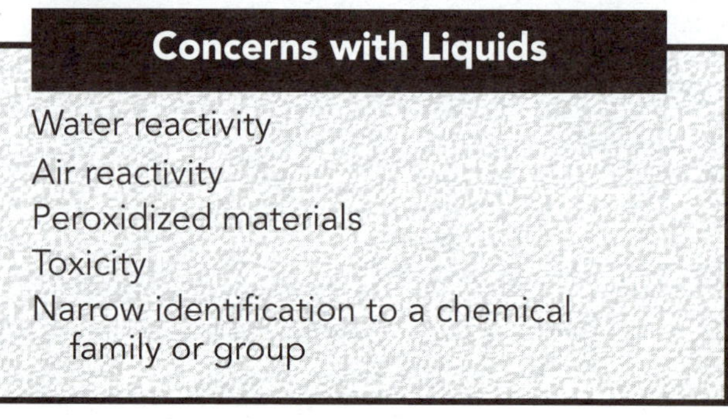

Concerns with Liquids
Water reactivity
Air reactivity
Peroxidized materials
Toxicity
Narrow identification to a chemical family or group

TABLE 9-3

HazCat Solid Sampling		
Test	**Instructions**	**Results**
Water reactivity	Place a pea-size sample on the watch glass. With a pipette place two to three drops of neutral water on the sample.	A lot of bubbling indicates severe water reactivity; a little bubbling indicates slight water reactivity.
Cyanide test	Add a pea-size amount of the solid material to ½ inch of water in a test tube. In another test tube place 1¼ inch of iron citrate and a pinch of ferrous ammonium sulfite. Mix the two tubes together, pouring the second tube's contents into the first. Shake this mixture for at least a minute. Add five to ten drops of hydrochloric acid (3N) to the mixture.	A color change to dark Prussian blue is an indicator for cyanide.
Oxidizer test	Place a pea-size sample on the watch glass, add one to two drops of hydrochloric acid (3N) to the oxidizer test paper and then dip the paper into the sample. [*Author's note:* The HazCat kit instructs you to add the acid to the unidentified for some tests after this initial test. My preference is to add the acid to the unidentified then test with the test paper, and then continue down the chart with the other HazCat tests.]	If the paper turns blackish blue, blue, purple, or black, an oxidizer is present. [*Author's note:* The HazCat system has several other tests to run to confirm the presence of an oxidizer. It is recommended that you continue with those tests.]
Peroxide test	Place a pea-size sample of the unidentified material in a watch glass and add four to six drops of water, so that the sample is a slurrylike solution. Dip the peroxidizer paper into the sample.	Wait up to 15 seconds, look for a color change from white to blue, and refer to the package for the concentration of the sample.
Sugar test	*Warning:* Be careful where you point the test tube while heating this solution. In one test tube mix ¼ inch of copper sulfate with a ¼ inch of alkaline tartrate. Stopper and shake the mixture until it becomes a rich blue color. In another test tube filled with ½ inch of water, add half of a pea-sized amount of the unidentified. If you have a liquid and you want to test it for sugar place a ¼ inch of the liquid into a test tube and follow the remaining instructions. Add 1 drop of 3N hydrochloric acid to the test tube containing the unidentified. Over a torch flame gently heat the acidified unidentified to almost boiling. Be careful when heating liquids in a test tube, as they can easily boil out. Add equal amounts of the blue sugar test to the heat mixture of the unidentified. Try not to fill the test tube to over half full at this time. If there is no orange to copper precipitate, gently heat the mixture some more.	A color change to yellow/orange then to red/brown precipitate indicates sugar. If the solution starts to turn orange or yellow then turns clear green, too much acid was added to the unidentified or the unidentified was too acidic. Flour and starch are complex sugars and may give a slight reaction. If the unidentified forms either a suspension or a gel, perform the flour test.
Flour test	Place a pea-size amount of the unidentified in a watch glass. Add two to three drops of potassium iodide to watch glass.	Color change to orange to blue/black indicates starch or flour.

Table 9-4 provides just a few of the many tests that can be accomplished by the HazCat kit for liquid materials. The HazCat system has full flow charts and provides extensive documentation as well as a process for determining the identity of unknown liquids. Refer to the HazCat kit and instructions for more information. These tests are provided to you as they give a quick overview of some of the initial characterizations listed in Table 9-2.

TEST STRIPS

One of the easiest multiple chemicals tests that exists for street responders are chemical test strips, as shown in Figure 9-7. These strips come in two forms: a **chemical classifyer** (Spilfyter™) and a **wastewater** version, both produced by JV Manufacturing. Although there is some duplication between the two strips, responders should have both available. Much like pH paper, these strips are for testing

TABLE 9-4

Liquid Sampling

Test	Instructions	Results
Water reactivity	Place a small amount (¼ to ½ inch) of the unidentified liquid in a test tube. With a pipette with neutral water, place two to three drops into the test tube.	A lot of bubbling indicates severe water reactivity; a little bubbling indicates slight water reactivity.
Specific gravity/polarity	Put ¼ to ½ inch of water in a 40 ml vial, then add four to six drops of the sample to the vial.	If the material dissolves in the water and mixes with the water it is polar (water soluble). If the material does not mix with the water, it is considered nonpolar. If the sample floats on the water, it has a specific gravity less than 1. If the material sinks in water, it has a specific gravity of greater than 1.
Flame test	Take the unidentified sample in a watch glass and light a match, preferably on a pair of forceps. Starting at 6 to 8 inches above the sample, slowly lower the match to the sample, continuing until you reach the sample. If it has not ignited, touch the match to the sample.	If the sample ignites at a distance of 2 inches and above, the sample is extremely flammable. If the sample ignites from just above the sample to 2 inches, the sample is flammable. If the material ignites when the match touches the surface or is slightly above and ignites after heating a few seconds, the material is combustible. If the match increases in flame height when touching the surface, the material is slightly combustible. If the match is extinguished when it touches the surface of the material, it is noncombustible.
Oxidizer test	Place four to six drops of the sample in the watch glass and add one to two drops of hydrochloric acid (3N) to the oxidizer test paper. Then touch the paper to the unidentified. [*Author's note:* The HazCat kit instructs you to add the acid to the unidentified for some tests after this initial test. My preference is to add the acid to the unidentified and then test with the test paper. Then continue down the chart with the other HazCat tests.]	If the paper turns blackish blue, blue, purple, or black, an oxidizer is present. It is important to watch for the results immediately as after time the paper will change color due to the oxygen in the air. [*Author's note:* The HazCat system has several other tests to run to confirm the presence of an oxidizer. It is recommended that you continue with those tests.]
Peroxide test	Dip peroxidizer test paper that has been wetted with water into the sample. For hydrocarbons, dip paper into sample and allow the liquid to evaporate. Then dip the test strip into water.	Wait up to 15 seconds and look for a color change to blue. Refer to the package for the concentration of the sample.

TABLE 9-4 *(Continued)*

Liquid Sampling		
Test	**Instructions**	**Results**
Hydrocarbon characterization (iodine chip test)	Place ½ inch of the sample in a test tube. Add one or two iodine chips to the test tube. Some slight agitation may be necessary. The second part of the test, after the color change, is to add some distilled water to the test tube.	Red (burgundy) indicates an unsaturated hydrocarbon. If the unidentified floats when several drops of water are added to the test tube, it is most likely benzene or toluene. If the unidentified sinks in water it is *probably* a halogenated hydrocarbon. The most common halogenated hydrocarbons are perchloroethylene, trichloroethylene, and chlorobenzene. If you add water and the top is burgundy/red and the bottom is yellow or clear then BTEX (benzene, toluene, ethylbenzene, and xylenes) and polar liquids are indicated. Purple indicates a saturated hydrocarbon. A purple indication is only possible when there is no functional group except a halogen on a saturated hydrocarbon (no double bonds). If the unidentified floats when water is added it is most likely paint thinner, naptha, mineral spirits, or stoddard solvent. If the unidentified is slightly oily, it is kerosene. Siloxanes are rare products that float and produce a purple color change. If the unidentified sinks when water is added, it is either carbon disulfide or a high molecular weight hydrocarbon. Carbon disulfide is *highly volatile*, extremely flammable, toxic, and smells like rotten pumpkins. The most commonly encountered halogenated hydrocarbons are carbon tetrachloride and methylene chloride. If you add water and the top is purple and the bottom changes to yellow or clear, then paint thinner, mineral spirits, kerosene, or jet fuel is indicated. Yellow or light orange indicates polar hydrocarbons such as acetates, alcohols, ethers, esters, nitriles, acids, or a mixture. Orange is predominant when there are two or more characteristics. If after water is added the mixture separates and produces a new color, a mixture is indicated. [*Author's Safety Note:* If the material also has a pH of >10 dispose of the material carefully as it may be ammonia, and the addition of iodine and ammonia forms nitrogen tri-iodide, a material that may explode if it is allowed to dry and is then disturbed. It is shock sensitive. PPE is important when sampling.]

(continues)

TABLE 9-4 (*Continued*)

Liquid Sampling		
Test	**Instructions**	**Results**
Hydrocarbon characterization (iodine chip test) (*continued*)	Place ½ inch of the sample in a test tube. Add one or two iodine chips to the test tube. Some slight agitation may be necessary. The second part of the test, after the color change, is to add some distilled water to the test tube.	Red/orange indicates a double bond and a polar group such as phenol. Brown or opaque is an indication of gasoline. Add water to the solution. The solution should separate, and the bottom should be yellow, and the top will remain brown. This change also indicates the presence of gasoline. A brown color also may indicate a mixture of hydrocarbons. Gasoline turns brownish black and stains the sample jar in few minutes with black spots and a black ring around the liquid level. Diesel fuel also turns muddy brown, but the black spots begin to appear after 1 to 2 hours, and the ring around the liquid level does not appear for 6 to 12 hours, although the bottom of the sample jar may be slightly brown/black after a few minutes. Unidentified with a pH near 11 or above 11 reacts, smokes and remains indicates amines, guanidines, pyridines, or other nitrogenated organic compounds; hydrazine salts or weak hydrazine solutions. Look for copious amounts of purple smoke. Unidentified with a pH near 7 reacts, smokes, and remains (or changes to dark red) indicates turpentine. Unidentified reacts, liquid boils and becomes clear indicates a triple bond. Unidentified immediately catches fire indicates hydrazine. No color change and the viscosity of kerosene indicates a carbon chain of seventeen carbons or more.

Source: Excerpted with permission from the *HazCat Abridged Manual for Field Use 1995* by Robert Turkington. HazTech Systems Inc. of San Francisco, California.

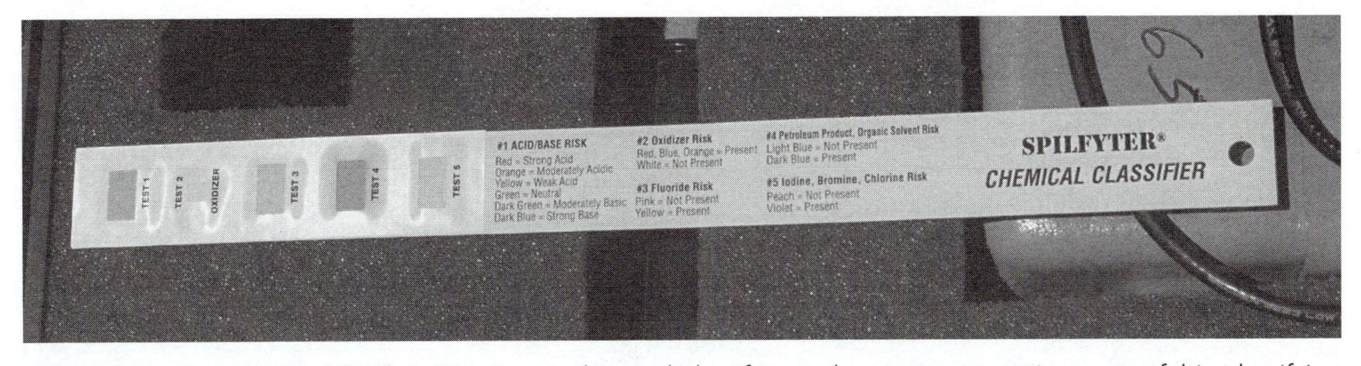

Figure 9-7 Two types of Spilfyter™ strips, a chemical classifyer and a wastewater strip, are useful in classifying unidentified substances.

unidentified liquid samples and can assist in identifying unknown materials. There are five tests on each strip, and the manufacturer intends that the strip be dipped into the unidentified material. However, it is recommended that a sample be taken from the unidentified using a pipette and placing a small drop onto each test area as shown in Figure 9-8. Some useful interpretation can be done if the sampling is done in that way. Pure compounds initially produce more vibrant color changes, while mixtures or less concentrated chemicals produce less vibrant color changes and usually take a little longer. If the strips were placed into an unidentified chemical that was slightly thick and had color to it, a color change is more difficult to see.

The Spilfyter™ tests for pH; oxidizer, fluoride, and chlorine/iodine/bromine risks; and hydrocarbons. The oxidizer risk test is important as it is a quick test for potentially explosive materials. The test for fluoride assists in the identification of hydrofluoric acid (HF). If you get a pH of 0–1, are not sure what type of acid is present, and the test for fluorine is positive, then most likely the acid is hydrofluoric. Only a couple of fluorinated acids exist, with the most common being HF. The wastewater strip samples for pH, hydrocarbons, nitrite risk, and lead risk. One important test on the wastewater strip is the test for lead, which is commonly dumped as part of old paint waste. With these strips we can very quickly identify the chemical family that a material comes from and make some effective and safe response decisions. These test strips are a good lead-in for the tests found in Table 9-4. One good thing that these strips can do is rule out what kinds of chemicals are not present.

NOTE One good thing that these strips can do is rule out what kinds of chemicals are not present.

Other test strips can assist in the identifications of unknown materials. The most common test for peroxides, chlorides, fluorides, and other materials. Many of these require refrigeration and are quite

Figure 9-8 Although the instructions recommend that you dip the whole strip into the unidentified liquid, it is best to place a drop of the sample onto the individual tests. In this way a closer observation of the results can be accomplished.

TABLE 9-5

Spilfyter™ Chemical Test Strips	
Test	**Sensitivity**
Chemical Classifyer Strip	
pH	0–13
Oxidizer	1 mg/l hydrogen peroxide
Petroleum hydrocarbon	10 mg/l gasoline
Chlorine, bromine, and iodine	1 mg/l chlorine
Fluoride risk	20 mg/l fluoride
Wastewater Strip	
pH	0–13
Nitrites	1 ppm nitrite
Petroleum hydrocarbons	10 mg/l gasoline
Lead	20 mg/l (20 ppm) lead
Hydrogen sulfide	10 ppm

expensive. A section on the HazCat kit provides a description of the common test strips that are useful for emergency responders. Some other chemical-specific tests include those for lead and for polychlorinated biphenyl (PCB) levels. The lead tests involve the wiping of a test material suspected of containing lead across the object. If lead is there, then the indicator will change color indicating its presence. The PCB test involves mixing the sample with some test chemicals and looking for a color change. These test kits are available in a variety of test ranges and are relatively easy to use. Many private electrical transformers exist that have PCB oil in them above 50 ppm, although they have largely been removed from public service utility use.

One test kit that has both standard HAZMAT and terrorism response use are pesticide test tickets. The test kits produced by Neogen corporation are extremely sensitive devices and easy to use. They can detect down to the parts per billion range for organophosphates, thiophosphates, and carbamate pesticides. Essentially any material that is a cho-

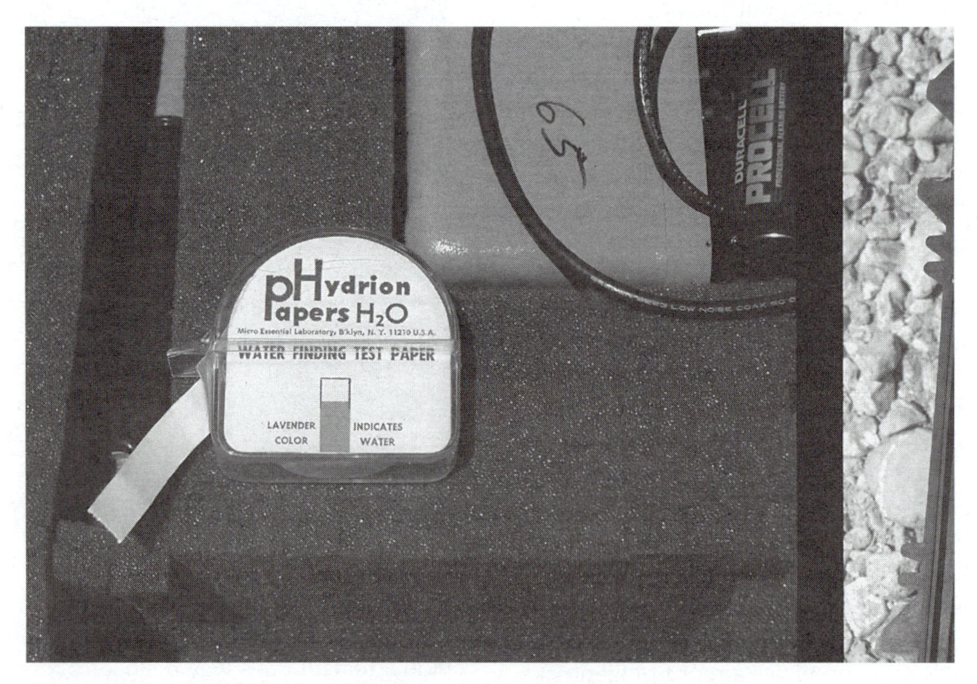

Figure 9-9 Water-finding test paper.

linesterase inhibitor will be picked up by this test, which includes the military warfare nerve agents. The only interferences are pHs less than 3 or greater than 8 or materials that use dyes (chromaphoric material). The most important point to remember in using this kit is to make sure the glassware and tools are cleaned with a bleach solution to eliminate false positives. Any residual pesticide will cause the next test to be positive.

WATER-FINDING TEST PAPER

The ability to test for water is important in the identification of unknown liquids. There are a couple of methods of identifying the presence of water, and the easiest is water-finding test paper as shown in Figure 9-9. This paper changes from white to violet in the presence of pure water. When the paper changes to any color there is water in the sample. The fact that the paper does not change colors does not mean water is absent, as some corrosives will bleach out the indicating dyes. Most corrosives have some amount of water in them, so if you get an extremely high or low pH, the water test may not be conclusive. A better method to check for water content is to break up some Alka-Seltzer™ tablets and put a pea-size amount in a test tube. Bubbling indicates the presence of water. In order to be assured that this test is successful, open the foil and break up the tablet just before the test, to avoid its breakdown by exposure to humid air. Another method to check for water is water-finding paste, commonly used by gas station owners to check for water in their storage tanks. The paste is spread on the dipstick and indicates when water is present.

DRUM SAMPLING

HazCat and Spillfyter™ strips come in handy in drum sampling. Abandoned drums or what is commonly referred to as a drum dump (see Figure 9-10) is fairly common across the country. The dumping of drums is a serious environmental crime and should be handled as such. One of the issues with drum dumps is how they are handled. The chances of someone ensuring compatibility and proper packaging are pretty slim whenever they dump the drums off the side of the road or wherever. Responders should leave abandoned drums in place until they have been sampled and the contents fully characterized. A smart dumper uses a transmission fluid

Figure 9-10 A typical drum dump.

drum, labeled as such and fills the drum with waste material, which to make the point we will say is a heavy metal contaminated dilute sulfuric acid. On the top of the sulfuric acid the dumper places a quart or so of transmission fluid, which will float on top of the acid. The dumper then finds a suitable location to dump the drum. The more time they have to get away the less likely they are to get caught. If the dumper chose a library in the downtown area to put this drum, eventually someone would phone some authority about the drum and someone would investigate the drum. By looking at the drum they would see the transmission fluid label or other nonhazardous markings and if they opened the drum they would only see the transmission fluid floating on top. Most responders would be inclined to call this a drum of transmission fluid, and leave the drum for later pickup. All the while the sulfuric acid is eating away at the steel drum, and generating hydrogen gas. At some point the drum is going to fail, either due to increased pressure or simple corrosion, releasing the metal contaminated sulfuric

CASE STUDY

Rock-n-roll drums, or drums that contain materials that are reacting, are an infrequent occurrence, but they do happen so you should plan for them. The most important consideration is isolation of the event, as the drum could rocket a considerable distance if it were to rupture or explode. The second consideration is to determine what the contents are. In the past two events that Baltimore County's team handled, the mixture in the drums was ink and cleaning fluids, and the others were paint and paint waste. In one case there was a single 55-gallon drum rocking and rolling on a loading dock. The facility backed up to a neighborhood, so letting the event run its course was not an option. We consulted with our chemist who advised a cocktail mixture of aluminum powder and alkaline waste to stop the reaction. Two HAZMAT team members approached the drum, and while one held the drum, the other opened it, slowly venting the pressure. Once opened, the cocktail mixture was added, and the reaction stopped. In the other incident two drums out of thirty were rocking and rolling, so our chemist determined a different cocktail mixture and all the drums were treated. If remote opening is possible that is always recommended, but speed is of the essence in this high risk mitigation, and having immediate access to a chemist is essential.

acid. One important issue is that most responders would assume it was actually transmission fluid that was in the drum, thus it is unlikely it would be isolated or picked up quickly. It would be a low priority even though the whole time that drum sits there its potential for a catastrophic release is increasing. Sampling the drum with a drum thief, which is a drum-sized straw, would take about a minute and would provide valuable information. Taking a sample and characterizing, in addition to HazCatting it would quickly indicate a mixture and that the bottom layer was a strong acid. At a minimum, the drum would have to be transferred or overpacked into an appropriate container and the hazard controlled.

SAFETY Responders should leave abandoned drums in place until they have been sampled and the contents fully characterized.

Opening drums presents a risk, but most drums that are going to be extremely risky to deal with provide several clues that there is a problem. Drums that are swelled, bowed up, expanded, or obviously under pressure present additional risk, and other methods of opening, such as remote opening should be considered.

SAFETY Drums that are swelled, bowed up, expanded, or obviously under pressure present additional risk, and other methods of opening, such as remote opening should be considered.

Most drums that have been sitting outside have some pressure. Unscrewing the bung cap very slowly, a quarter inch at a time, will safely vent the pressure. While doing so, the air monitors should be close to the bung, to evaluate the released gas. Most HAZMAT teams use poly drum thiefs for their sampling as glass thiefs have a tendency to break. If you are sampling for evidence you must use glass, and they are only used once. Most teams that are collecting evidence use coliwasa tubes, as they can give an exact cross-sectional representation of what is in that drum. There are both poly and glass coliwasa tubes. Anything used for evidence must be certified clean, kept sealed, then used one time to collect a sample. Once the sample is collected, it can be tested the same as any other unidentified liquid.

INFRARED THERMOMETERS

Infrared thermometers are useful in drum dumps to determine the temperature of the materials in the drum and to determine the liquid levels. Guided by a laser beam, the thermometers can determine the temperature of a surface from a considerable distance away. The exact distance is determined by the size of the target spot to be read. The target spot at 6 feet is 2.4 inches with the Raytech S18 model. The temperature range that these thermometers can read is –25° to 1000°F. Many fire departments now use thermal imaging devices to detect heat, but some thermal imaging devices cannot provide the temperature, only a visual indication that the spot being read is a higher temperature than the adjacent area.

SUMMARY

Detection devices such as mercury detectors, HazCat kits, and test strips are a valuable resource for identifying unknown materials. Both low cost and expensive technical devices are available to responders. Some initial detection can be done with the low cost devices, and then further characterization can be done outside the hot zone with the more technical kits. Developing a system that combines many different technologies can be difficult and Chapter 10 can assist in the development of a sampling system.

KEY TERMS

Chemical classifyer A testing strip that includes five tests; a companion test strip is the wastewater strip.

HazCat™ kit A chemical identification kit that can be used for solids, liquids, and gases.

Wastewater strip A strip that does some additional tests beyond the chemical classifyer, which see.

REFERENCE

Turkington, Robert. 1995. *HazCat Abridged Manual for Field Use,* Haztech Systems Inc., San Francisco, CA.

TACTICAL USE OF AIR MONITORS

- Introduction
- Risk-Based Response
- Basic Characterization
- Meter Response
- The Role of Air Monitors in Isolation and Evacuation
- Detection of Unidentified Materials and Sampling Priorities
- Summary
- Key Terms

INTRODUCTION

The tactical use of air monitors is by no means an easy skill, as there are many factors to consider including the occupancy and location, the type of material and its state of matter, spill location, weather, and finally the task to be completed. One of the best methods for learning how to effectively use air monitors is to use them for every incident, even when the preponderance of information tells you that you really do not need all the instruments or that instruments may not be required at all. It is a good habit to carry monitors when on calls that do not involve hazardous materials, so you can learn what they do or do not react to. The monitors can be used when you do company tours or inspections so you learn what you may expect to find in facilities under normal conditions. Knowing what kind of reactions and levels you can expect is invaluable when you arrive at a facility under emergency conditions. Suppose that the metal oxide detector in one chemical manufacturing facility reads 250 units. In itself that number does not mean anything other than it is 250 units out of 50,000 units, which could be considered a low reading. The problem is that an inexperienced responder on an emergency call may interpret 250 as part of the emergency condition and think that there is a problem when in fact there is no problem. A reading of higher than 250 for this plant would indicate a problem.

SAFETY One of the best methods for learning how to effectively use air monitors is to use them for every incident.

Note: This chapter on the tactical use of air monitors provides some general rules and examples of tactical deployment of air monitors. Each situation is different, and the responder should always wear a minimum level of protective clothing. Successful monitoring is based on regular calibrations, bump tests, and training, but experience also is a factor. Science is not black and white, it has many shades of gray. It is the ability of air monitors to detect these gray situations that will keep you safe.

RISK-BASED RESPONSE

The best use of monitors is in a risk-based response (RBR) profile, in which the monitors assist in the risk assessment of the situation. The monitors are your safety mechanism, much like a bulletproof vest that protects a law enforcement officer. The benefit of RBR is improved safety for responders and quick assessment, identification, and mitigation. By using improved air-monitoring skills, appropriate decisions can be made about PPE, isolation and evacuations, and the severity of the event. With RBR these decisions can be made within minutes, usually less than 5 minutes for a well-trained crew. However, RBR requires educated and experienced personnel using air monitors and interpreting the results. Air monitors are dumb devices; they require human intervention to make the right decision based on what the monitor reports. The first barrier that needs to be hurdled is that responders need to trust the instrument. This trust is based on the hope that the instrument is calibrated regularly and the responder understands the true meaning of the interpreted results. A properly calibrated instrument does not lie; it is up to the responder to confirm the identity of the material and to interpret the numbers.

The basic premise behind RBR is the use of air monitors to classify or characterize a chemical into one or more risk categories. The three risk categories are fire, corrosive, and toxic, and once a risk category is identified crews can be properly protected. The RBR works well when dealing with an unidentified chemical or a potential release situation. HAZMAT teams should avoid the use of the term *unknown* as it really does not exist. HAZMAT teams do not respond to unknowns, as there is always something known that is available. People do not dial 911 just to get the HAZMAT team searching in some building looking for "something." There is a reason someone called 911, and that is known information. Granted, the information may be limited or the exact identification of the released substance may not be known, but every response has some "known" factors. HAZMAT crews need to capitalize on those factors and use them for their benefit. Using basic chemistry such as the state of matter assists us in the RBR process. A release of an unidentified solid material, is not a major event, as a solid does not present an overt hazard unless you eat it or touch it. In the investigative phase, the level of PPE can be relatively low, as the risk is low. In the cleanup phase in which there is potential for extensive contamination, then PPE may be bumped up, but the tasks are different. No one chemical suit is best for HAZMAT situations. The risk category, chemical and physical properties, and the task determine the best type of PPE that should be worn. It is a misconception that Level A offers the best or the highest amount of chemical protection. The hazards from the suit itself, such as heat stress, limited visibility, mobility, and communication, are all major life safety issues. HAZMAT injuries are from heat stress, not chemical exposures. There is great potential for fatalities if you believe that Level A is the safest suit. Level A suits are appropriate for certain tasks, such as dealing with corrosive gases above the **immediately dangerous to life or health (IDLH)** level, or in toxic-by-skin-absorption IDLH atmospheres. Level A is certaintly appropriate when there is potential for whole body contact, or when responders may become covered with a material that may be corrosive or toxic. No matter what the toxicity, Level A is not appropriate for any situation in which there is a fire risk, and the flashover garments provide a false sense of protection. Remove the fire hazard, then wear appropriate PPE. When dealing with a fire risk, wear appropriate fire protective clothing.

When using a RBR profile you need a minimum of three detection devices, which includes pH paper, an LEL monitor, and a photoionization detector. Using these three devices you are covering the corrosive, fire, and toxicity risk categories. When the LEL monitor is discussed it is assuming that a three- or four-gas instrument is being used, which means that you are also sampling for oxygen content, CO, and H_2S. At this point a discussion of PPE to do this investigation is appropriate. Firefighters protective clothing and SCBA is the recommended protection level, but common sense must prevail. Review your HAZMAT calls, and see which ones truly needed Level A protective clothing. The HAZWOPER regulation (29 CFR 1910.120) mandates the use of Level A in situations in which a material that is toxic by skin absorption is found above the IDLH level. The regulation also states that Level B is acceptable for unidentified situations. The regulation does not mandate a level of protective clothing, with the exception of the conditions provided for Level A. There are some conditions in which Level A is recommended, which include confined spaces and highly toxic environments. A situation in which there are fatalities in a building, is a probable candidate for Level A, as well as situations in which extensive contact is possible. We frequently encounter above IDLH conditions, including incidents with fatalities involving carbon monoxide and we use turnout gear (TOG) and SCBA, which is appropriate for the situation. Using Level B for unidentified materials is recommended in HAZWOPER. One item to consider though is what is the risk of the commonly encountered chemical? Most HAZMAT teams respond to flammables and combustibles, usually in excess of 50 percent of the total number of responses. By choosing TOG and SCBA one is protected against the chemical most likely to be found. In both the TOG and the Level B, the SCBA protects against toxicity. The only true risk that Level B does not protect against is corrosive vapors, which are easily detected. Let the task dictate the level of protective clothing, and when coming in contact with a chemical, wear appropriate chemical protective clothing (CPC). When investigating a possible release, wear SCBA and TOG, which is more than adequate.

BASIC CHARACTERIZATION

Basic characterization of the event is done by the RBR system, and some tactical decisions can be made by the interpretation of these results. Table 10-1 covers the top ten most frequently released chemicals and what can be expected with the three detection devices. Understanding how these chemicals respond to the monitors puts you closer to an identification. The table is divided into several areas and provides some additional clues to identification. In many cases there are not common materials that mimic the properties and readings of the chemicals listed.

METER RESPONSE

This section provides some insight into how a variety of chemicals may affect the range of air monitors. To classify an unidentified material it must first be classified into one of the three risk categories. Once classified, if it meets only one of the three risk categories, responders can work to specifically identify the material through a number of methods. If it meets more than one risk category, then identification may be more difficult depending on the circumstances. Although it is not often mentioned in these sampling protocols, make sure to always determine the background radiation and wear a rad pager or dosimeter.

Of all of the risk categories, the corrosive risk is the easiest to determine. The detection device is easy to use and to interpret. The range of corrosives on the street is limited to a few common materials. If using multirange pH paper, a color change to red indicates an acid being present. If the pH paper changes above or away from the spill, then the acid is a high vapor pressure material. The most common materials that are acidic and have a high vapor pressure are hydrochloric acid, hydrofluoric acid (HF), acetic acid, and **oleum.** The HF has a more characteristic vapor cloud and is more aggressive than hydrochloric acid. The oleum is the most aggressive as far as a vapor cloud and reaction with other materials, but does not indicate on the pH paper until you get fairly close to the spill. The higher the percentage of oleum, the farther away it will indicate. Many acids are low vapor pressure acids and only indicate when the pH paper is dipped into the liquid. Further tests are needed to narrow down the type of acid released. Common acids that are low in vapor pressure include sulfuric and phosphoric.

The catalytic bead LEL sensor only picks up flammable gases and those materials that are emitting flammable vapors. If a catalytic bead LEL sensor provides a reading of 1, then there is a flammable material released. If you cannot locate a liquid spill then the reading is from a gas. If you find a liquid spill and the catalytic bead LEL sensor is reading, and as you get closer to the release the meter starts to climb, then the spilled material is flammable. The amount of the reading helps determine how the meter reacts to the material. The quicker and higher the meter goes, the more flammable the material is. A metal oxide sensor (MOS) detects both flammable and combustible materials. The MOS also picks up inorganic materials that a photoionization detector may not. Keep in mind that the MOS will also pick up some particulates in the air as well. The benefit of a MOS is that it detects tiny amounts of many things, long before they become harmful, and in some cases the other detection devices may not pick up the material. A material that causes an MOS to rise rapidly and go above 1,000 is a flammable. The higher the number and the quicker the rise, the more flammable the material is.

The photoionization detector picks up organics and some inorganics. The most common material that a PID reacts very well to is solvents, all flammable liquids. The quicker and the higher the rise, the more flammable the material is. The PID is used to determine the presence of toxic risks, but many toxic risks also present a fire risk as well. Once a PID goes above 500, the material is usually picked up by a catalytic bead LEL sensor. The MOS LEL tracks much like the PID and rises accordingly. However, the MOS cannot replace a PID, as the MOS technology is way too erratic to be depended on for definitive toxicity readings. It does fine indicating the presence of a material that should not be in the air. It cannot be related to PPM and should not be used for this purpose. Once a PID picks up a material, you must use colorimetric tubes, and possibly iodine chips for liquids. The FID can be used at this point to determine if the material is organic or inorganic. If the FID picks up the material, it is an organic substance.

A common question is, what happens when a truckload of chemicals overturns and they all mix together? In reality, the mixing of the materials is not a large concern unless they react with each other.

TABLE 10-1

Top Ten Chemicals and Meter Responses

Meter Response	Ammonia	Sulfur Dioxide	Chlorine	Hydrochloric Acid	Propane
pH	>12, indicates in vapor; has a high VP	<2, indicates in vapor; has a high VP	<2, indicates in vapor; has a high VP	<1, indicates in vapor; has a high VP	No response
LEL	High levels in a building indicate a flammable atmosphere	No response	No response	No response	Yes
PID	Good response; a good detection device for ammonia	No response	No response w/10.6, but response with 11.7 eV lamp	No response	Only for 11.7 eV lamp
O_2	>5000 ppm will drop O_2 by 0.1%	>5000 ppm will drop O_2 by 0.1%	>5000 ppm will drop O_2 by 0.1%	>5000 ppm will drop O_2 by 0.1%	>5000 ppm will drop O_s by 0.1%
CO	No response	No response	No response	Usually no response may indicate on some instruments; usually accompanied by a H_2S reading	No response
H_2S	No response	No response	No response	Usually no response may indicate on some instru-ments; usually accompanied by a CO reading	No response
Odor	Strong, irritating, will be noticeable	Strong, irritating, will be noticeable	Strong, irritating will be noticeable	Strong, irritating will be noticeable	Distinctive odor
Next step in detection	Colorimetric tube to confirm	Colorimetric	Spillfyter strip (liquids) or colorimetric tube to confirm	Colorimetric	Colorimetric tubes

TABLE 10-1 *(continued)*

Top Ten Chemicals and Meter Responses				
Sodium Hydroxide	**Sulfuric Acid**	**Gasoline**	**Flammable Liquids**	**Combustible Liquids**
>12, indicates only in the liquid	<1, only in the liquid	Neutral (check leading edge)	Neutral (check leading edge)	Neutral
No response	No response	Yes	Yes	No response
No response	No response	Yes	Yes, usually a quick response	Responds with low readings, must be close to the liquid
No response	>5000 ppm will drop O_2 by 0.1%	>5000 ppm will drop O_2 by 0.1%	>5000 ppm will drop O_2 by 0.1%	>5000 ppm will drop O_2 by 0.1%
No response	No response	No response	No response	No response
No response	No response	No response	No response	No response
No response	No odor	Distinctive odor	Petroleum odor	No response
Colorimetrics	Colorimetric	Colorimetric tubes, iodine chip, or HazCat	Colorimetric, iodine chip, or HazCat	Colorimetric, iodine chip, or HazCat

Violent reactions usually occur prior to the arrival of responders, and any major event of this type is usually preceded by sufficient warning. By using a RBR philosophy it really does not matter what the mixture actually is. When chemicals mix, they still may present fire, corrosive, or toxic risks. Using standard RBR responders can be protected against all three hazards and can determine appropriate levels of protective clothing, isolation distances, and severity of the event. The exact nature of the mixture can be determined later and usually requires the use of a lab.

THE ROLE OF AIR MONITORS IN ISOLATION AND EVACUATIONS

The use of detection devices enables the IC to make immediate decisions regarding isolation and evacuation strategies. Following the recommendations in some reference texts or plume dispersion models in most cases results in too large an area being evacuated unnecessarily. The distances used in references are very conservative, and are set for a "normal" (worst-case) situation. By using the wide-ranging devices such as the LEL sensor and the PID, some initial isolation distances can be set. When readings begin to be encountered, this area becomes a potential isolation area (Figure 10-1). Isolation areas should be designated by hazard area, such as toxic, corrosive, flammable, explosive, collapse, and others. The designation should include the type of hazard that could be found in that area. Other devices such as the APD2000 or SAW MiniCad can be used for the same purpose (Figure 10-2). For the normal chemical releases the use of this tactical objective is easy, as the most common chemicals are easily detected. Using chlorine as the example, it would be easy to set up an isolation zone for a chlorine release. Some teams use the TLV or PEL or half that value to establish an isolation area. The IDLH would indicate the immediate hazard area, usually called the hot sector zone. All of the areas would be expanded to allow for wind changes, and product movement but real-time, on-site monitoring establishes the true conditions.

In some areas, especially major metropolitan districts, it is nearly impossible to effect an immediate evacuation. A high-rise building may have thousands of people working in it. A several block area may have hundreds of thousands of people in it working and living there. To say that we will evacuate this area quickly and safely is living in a vacuum. It is better to determine the hazardous environment and isolate only those areas that are affected or could become affected due to a shift in weather conditions or other change.

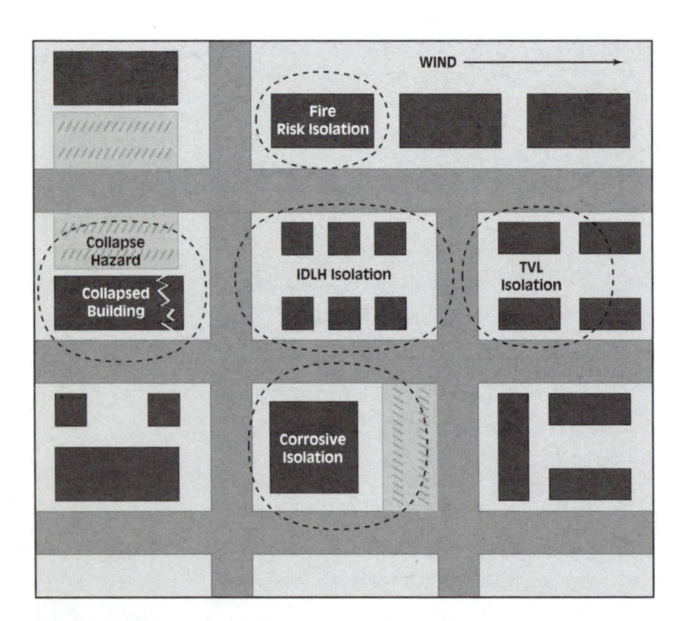

Figure 10-1 Isolation areas should be designated by hazard area, just not hot, warm, and cold.

Figure 10-2 Using an APD2000 to set up an isolation area.

DETECTION OF UNIDENTIFIED MATERIALS AND SAMPLING PRIORITIES

Figure 10-3 through Figure 10-8 are flow charts to guide you through the process of characterizing an unidentified material. These figures are an attempt to provide a starting point, upon which you need to expand. Local conditions dictate which methods may work best, and the flow charts assume that all the devices mentioned therein are available to responders. Some initial cautions must be provided to this section, as a 100 percent positive identification is nearly impossible in the field, and requires lab analysis. However, you can characterize an unidentified material to a minimum of a risk category. Many of the systems in place today to characterize a substance rely on one detection technology and do not take into account other technology available to a HAZMAT team. As with many things in the HAZMAT world, this information is what is normally found, and local conditions may change the way a chemical reacts, with weather being the predominant factor. One would not expect a chemical to act the same in Chicago in January as it would in Miami. The standard is 70°F, and you will need to calculate any temperature other than that. Once you gain some experience you may be able to skip some steps in these flow charts, as street experience plays a factor in characterization.

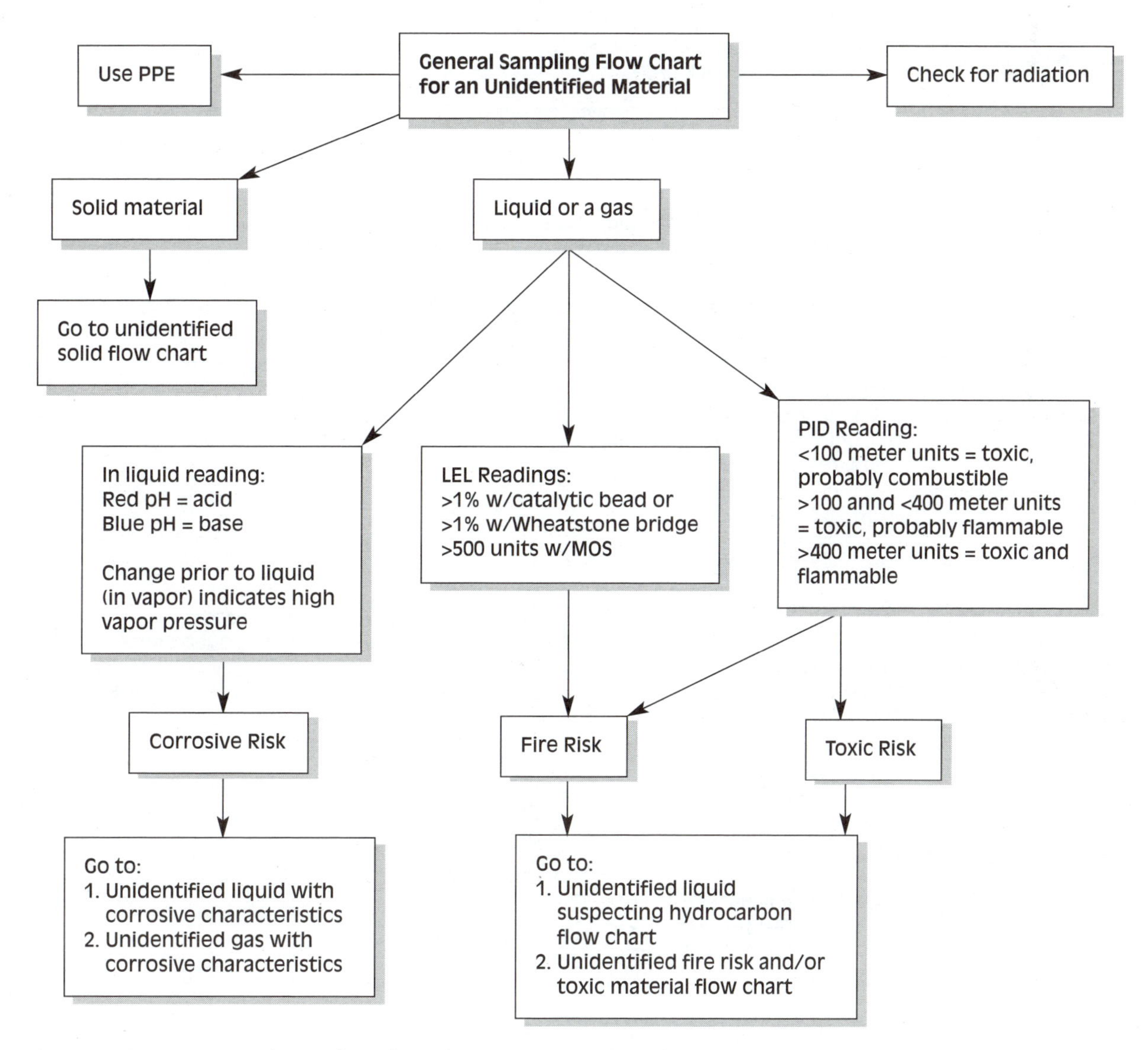

Figure 10-3 General sampling flow chart, to be used as the starting point.

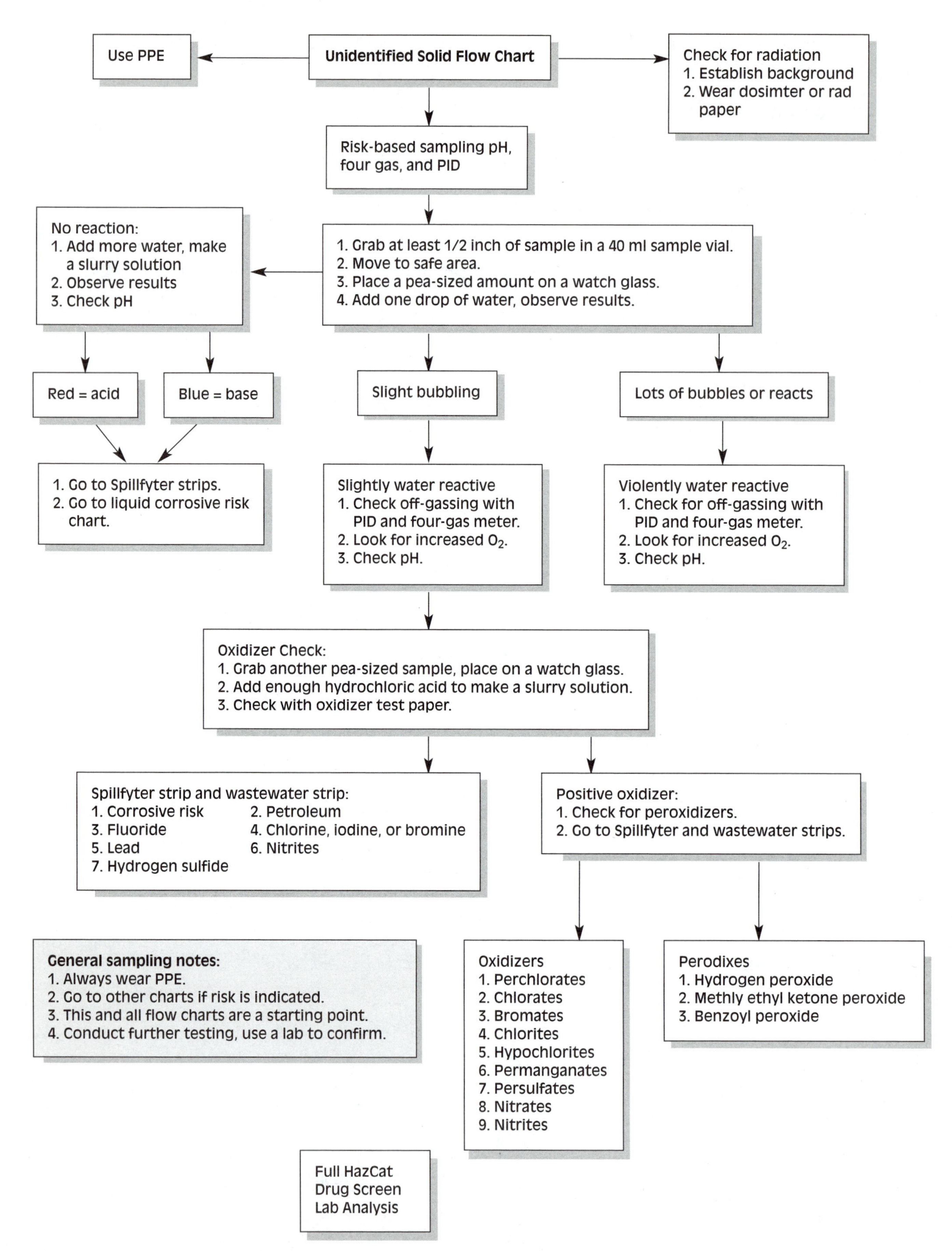

Figure 10-4 Sampling flow chart for unidentified solid materials.

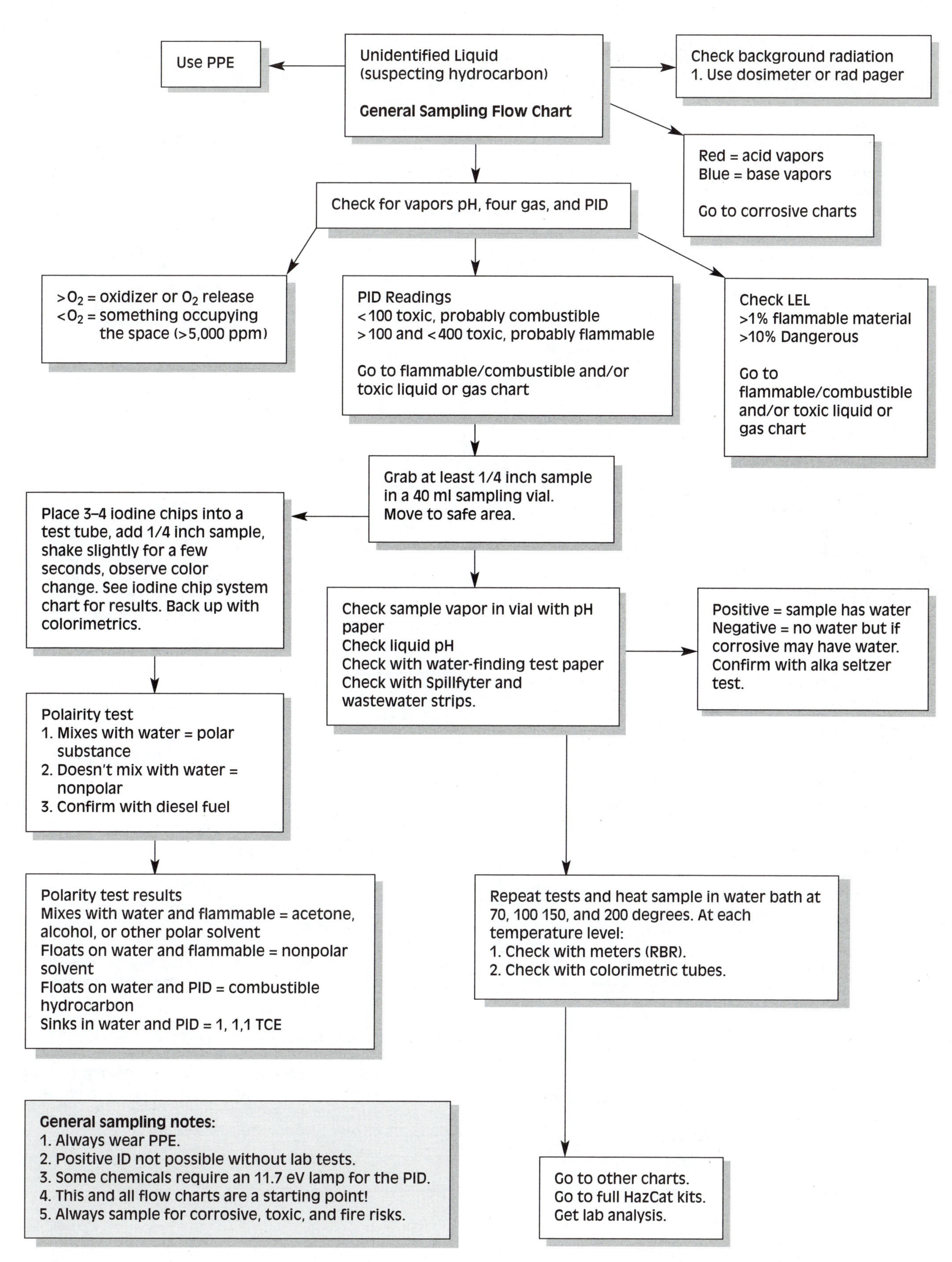

Figure 10-5 Sampling flow chart for unidentified liquid suspected of being a hydrocarbon.

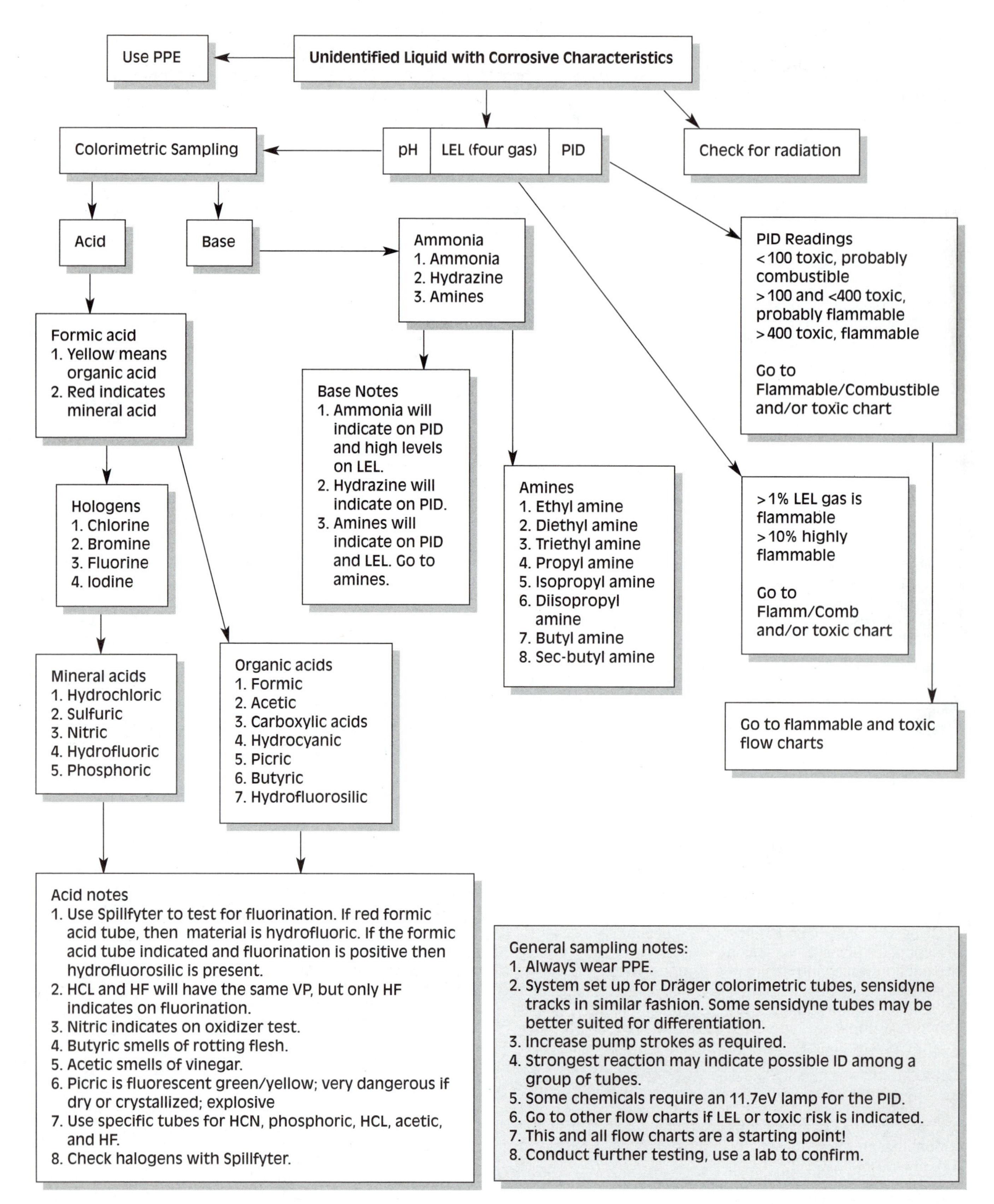

Figure 10-6 Sampling flow chart for an unidentified liquid with corrosive characteristics.

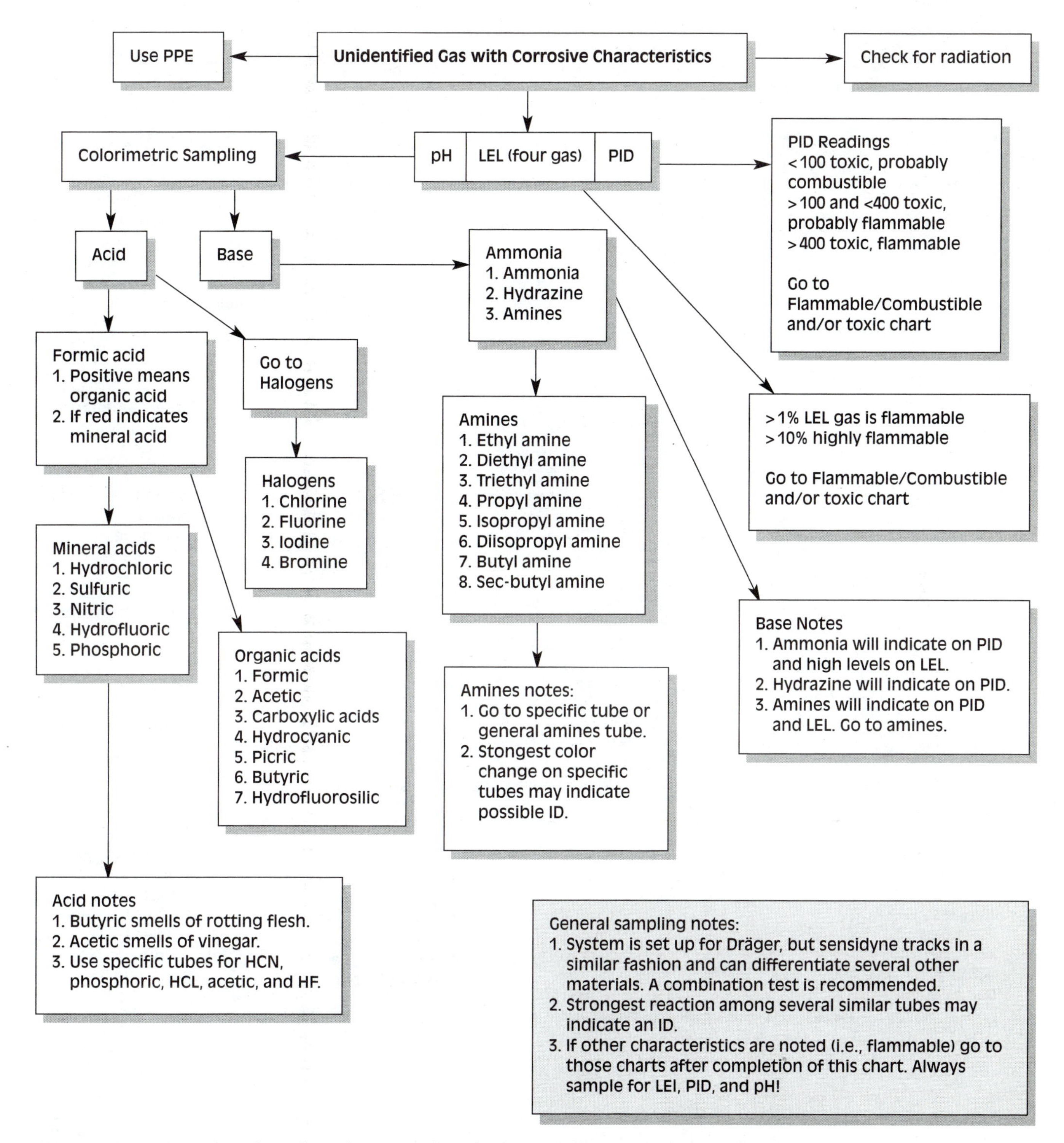

Figure 10-7 Sampling flow chart for an unidentified gas with corrosive characteristics.

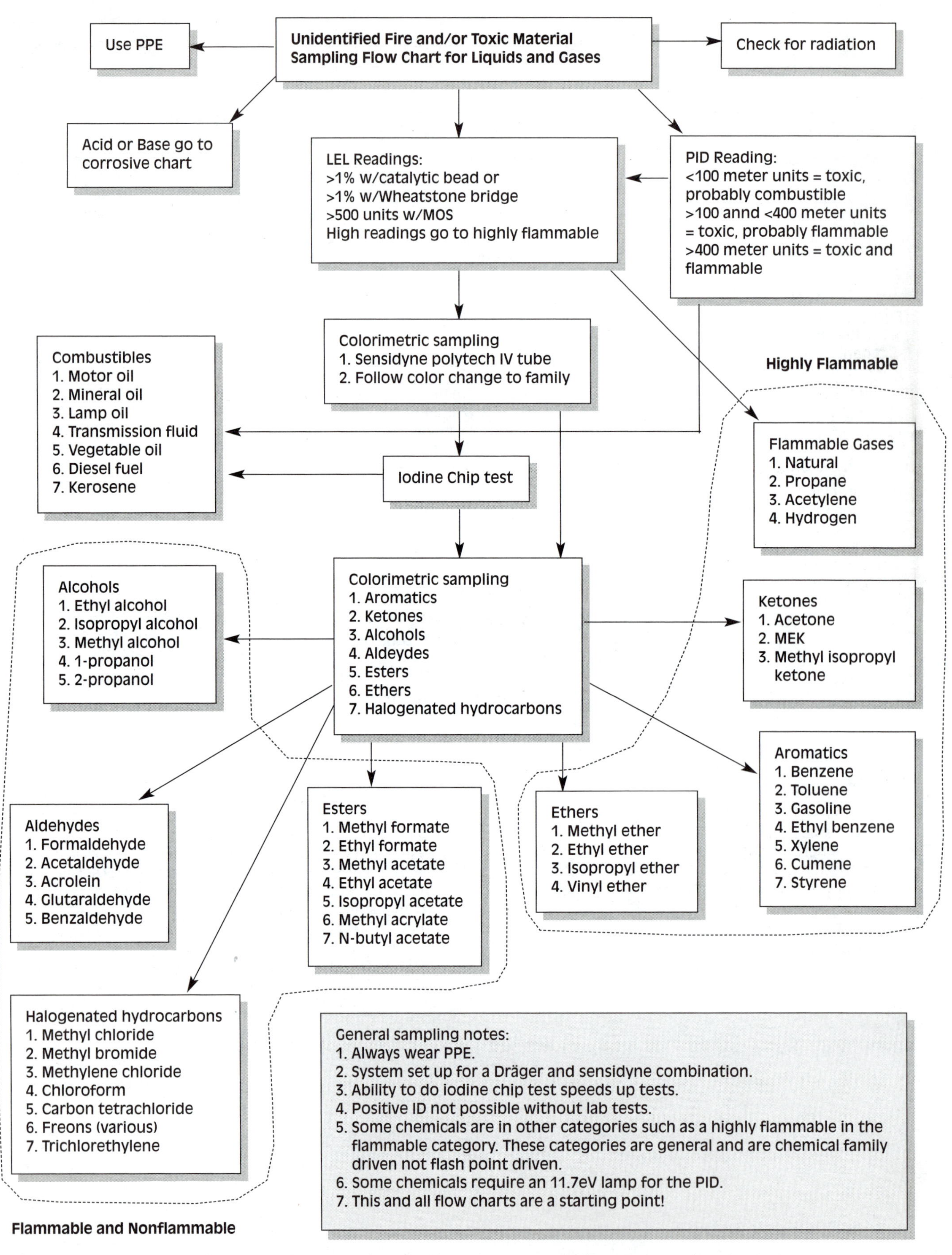

Figure 10-8 Sampling flow chart for an unidentified liquid or gas with fire or toxic risk indicated.

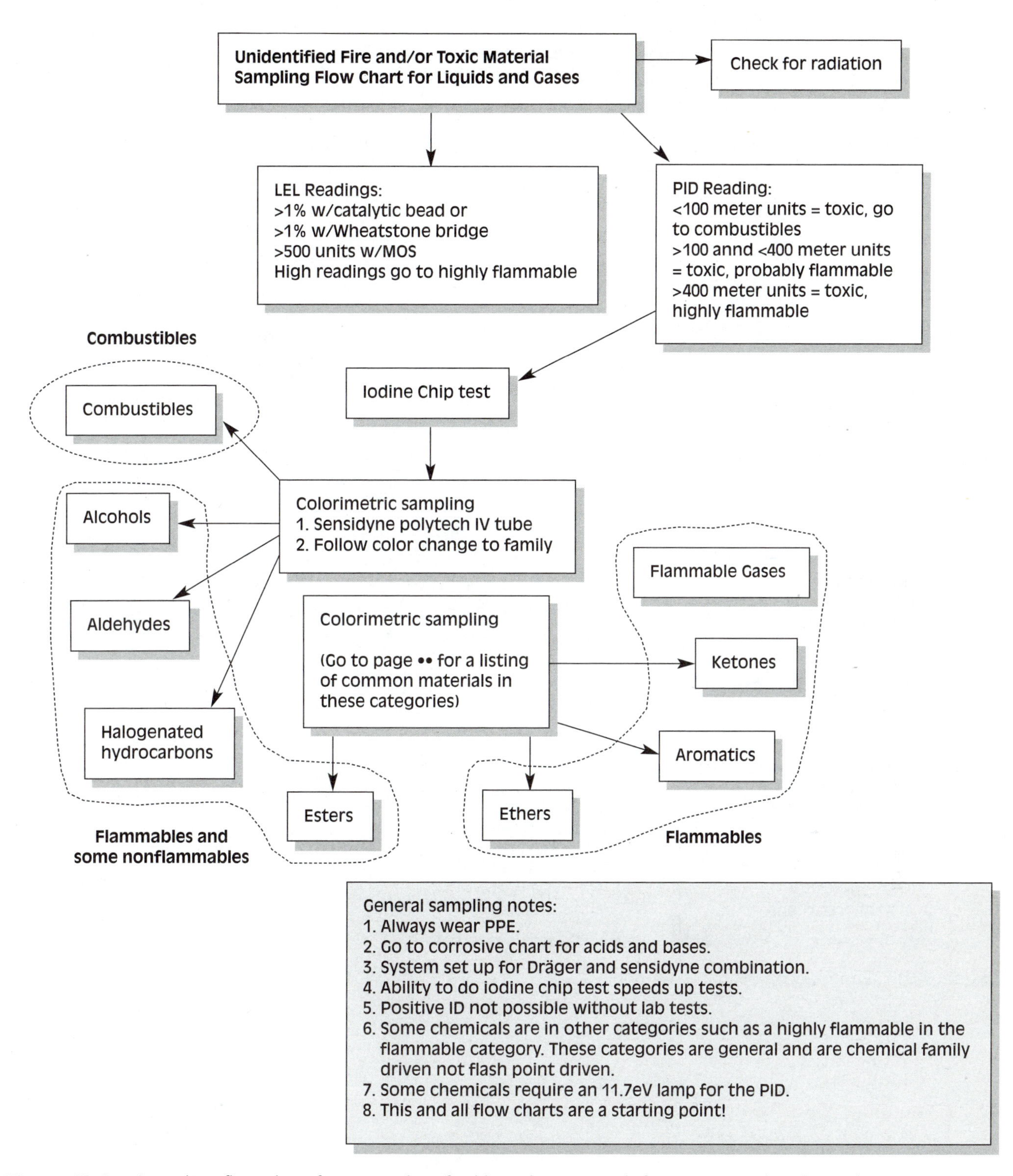

Figure 10-8 Sampling flow chart for an unidentified liquid or gas with fire or toxic risk indicated.

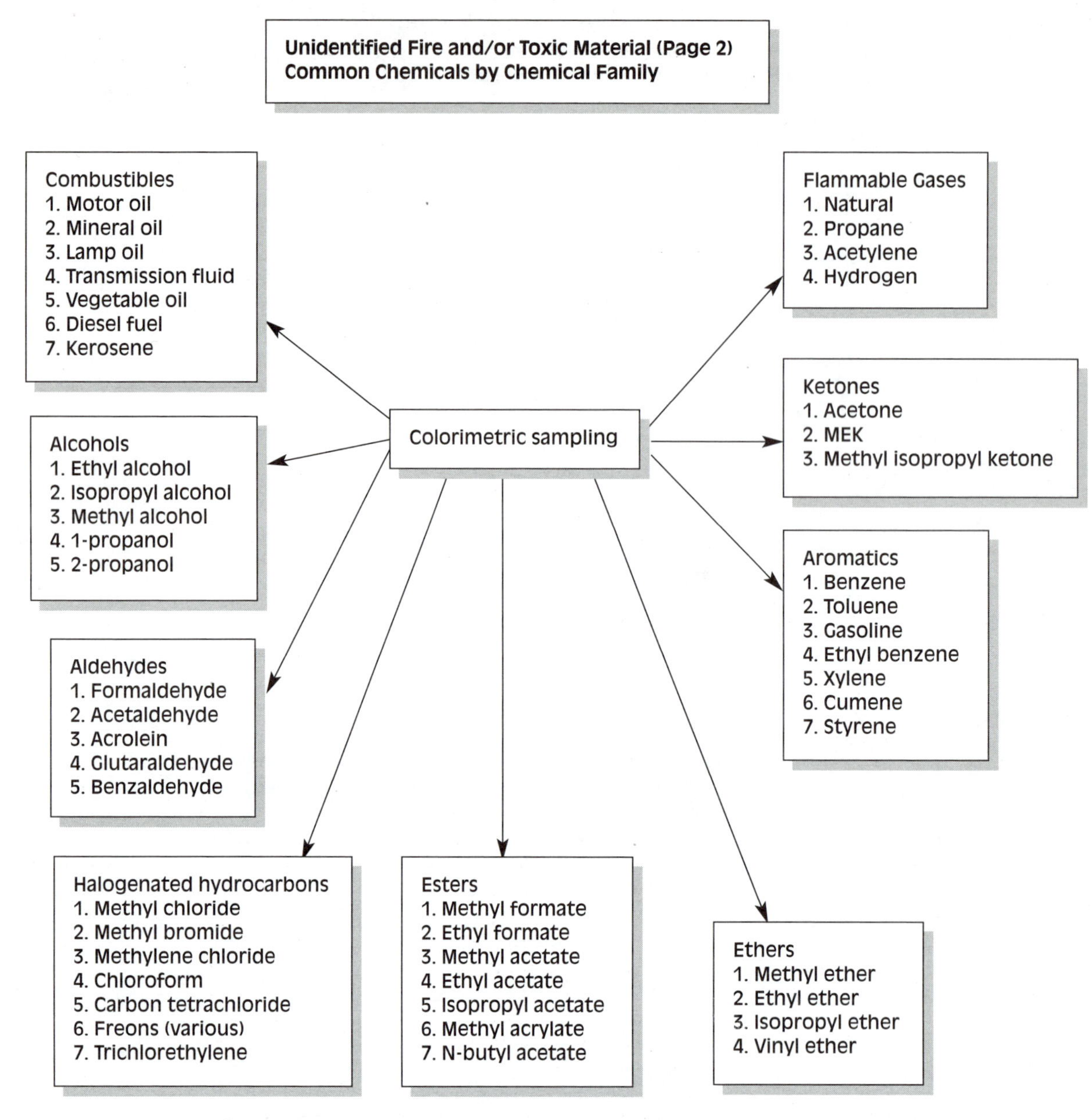

Figure 10-8 *(Continued)*

SUMMARY

The idea of risk-based response is one that benefits the responders, the community and anyone else who may be affected by a chemical release. Just on the basis of chemical characterization responders can gain valuable response information safely and quickly. The whole concept of identifying risk, evaluating the meter's response, and determining isolation and evacuation areas are important skills for the responder to learn. As that could best be accomplished in written form, there are guides provided to the reader to help with the evaluation process. Keep in mind these are guides, and your own knowledge and experiences play a factor in identifying an unknown material.

KEY TERMS

IDLH See *Immediately dangerous to life or health.*

Immediately dangerous to life or health Term used by OSHA to describe an exposure level in which a person is at serious risk for becoming unable to remove themselves from the affected atmosphere after 30 minutes exposure. Levels much above IDLH or exposures greater than 30 minutes can be fatal.

Oleum Concentrated sulfuric acid which has been saturated with sulfur trioxide.

TERRORISM AGENT DETECTION

- Introduction
- Methods of Detection
- Sampling Strategies for Terrorism Agents
- Warfare Agents Sampling Strategies
- Summary
- Key Terms

INTRODUCTION

With the increased concern with terrorism there is a need to be able to detect warfare agents, such as those provided in Table 11-1. Most HAZMAT response teams are using detection devices that were originally designed for military use and may or may not have been adapted to civilian use. The greatest concern with using military devices in the civilian world is that a considerable number of false positives are found with them. The military is less concerned with false positives, and in reality they do not impact their daily operation. The military does not mind having troops put on extra PPE for many hours, and the acceptable losses theory, in which the military calculates that a certain number of troops will be killed in an operation—something that the civilian responders find unacceptable—plays a factor here. The alarm levels also factor in the possibility of acceptable losses. One military detector alerts to the presence of vinegar, something that the military would not anticipate finding in the desert or the jungle. The military devices also are not designed to operate like their civilian counterparts. The screens are usually small and hard to read. The units typically only report back that there is a material present. The units may only report that a range of material is present without specifically providing an

TABLE 11-1

Warfare Agents

Agent Class	Common Agents	Military Designation*
Nerve	Sarin	GB
	Tabun	GA
	VX agent	V
	Soman	GD
Blister (Vesicants)	Sulfur mustard	H
	Distilled sulfur mustard	HD
	Lewisite	(L)
	Phosgene oxime	(CX)
Riot Control	Tear gas	CS
	Mace	CN
	Capsaicin (pepper)	CR

*Many detection devices still use the abbreviation for the military designation for the material.

exact level. When they do use a number to report back a reading it is just a number in range; the units are not parts per million or other common units of measure. There are no alarms or alerts to dangerous levels of chemicals that may be present. The units, however, are rugged, will withstand adverse weather conditions, and are easy to operate.

NOTE The greatest concern with using military devices in the civilian world is that a considerable number of false positives are found with them.

METHODS OF DETECTION

There are five basic methods of detection of warfare agents. The basic methods are:

- Test strips
- Colorimetric sampling
- Direct read instruments
- Military test kits
- Field/lab analysis

These methods mirror those for detecting dangerous substances in the civilian world. The direct read instruments, such as the **advanced portable detector 2000 (APD2000)** shown in Figure 11-1, provide readings much like the photoionization detector, although the technology to ionize the gas different.

Test Strips

Every HAZMAT team in the country should have a minimum method of detecting warfare agents. The lowest cost method is to use **M-8** (Figure 11-2) and **M-9** (Figure 11-3) **papers.** These two devices are easy to use and maintain; their only drawback is that they can only be used on liquids. These devices react to a large number of common chemicals and have a tendency to false positives.

NOTE Every HAZMAT team in the country should have a minimum method of detecting warfare agents.

M-8 paper is like pH test strips for warfare agents, and it indicates for nerve and blister agents. It provides a separate test for **VX,** and through a color change indicates which type of agent is present. The M-8 test kit comes with twenty-five test sheets, and a color indicating chart is provided in the front of the test booklet. A liquid sample is needed for the test, it requires at least 0.02 ml for the test to work, and has a thirty-second reaction time.

Figure 11-1 Using an APD2000 to look for potential terrorism agents at TOPOFF 2000, a large-scale terrorism exercise held in Portsmouth, N.H.

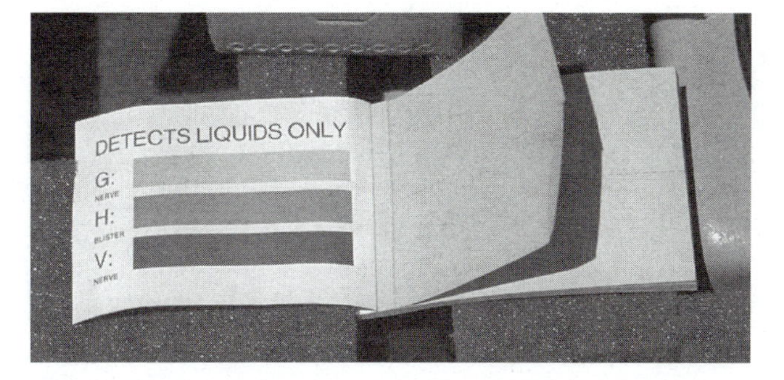

Figure 11-2 M-8 paper used to detect liquid nerve, blister, and VX agents.

Figure 11-3 M-9 tape used to detect liquid nerve and blister agents.

M-9 paper is better known as M-9 tape, as it is a roll of tape with an indicating layer on the outside surface. It is commonly placed on vehicles, boots, gloves, and other parts of the responder. M-9 indicates the presence of nerve and blister agents, through a red color change. M-9 does not differentiate between nerve or blister, it only indicates the presence of a warfare agent. M-8 and M-9 are used together as they both react to a differing set of false positives. M-9 is like M-8 and has a lot of false positives. M-9 is useful because it is more sensitive than M-8 paper, and will indicate first. In low levels there may not be a distinctive color change, but there may be pinpointed color changes on the tape. The pale green paper is impregnated with indicating dyes that turn pink, red, reddish brown, or red-purple after being exposed to liquid.

Colorimetric Sampling

One of the best methods for detecting warfare agents without using an electronic device is to use colorimetric sampling tubes (see Figure 11-4). Only a few companies (Dräger and MSA) make colorimetric tubes for specific warfare agents, but most companies provide tubes for the hydrolysis products of the agents. There are two primary methods in use for colorimetrics, a combination system that samples for a variety of agents simultaneously and sampling for individual agents independently of each other. Obviously with time being of the essence, the multisampling system has the advantage. The individual tests allow for more sampling flexibility and assist in the final identification, so a combination strategy should be employed. The multitest provides one shot at an identification, while the individual tubes can usually do up to ten tests. It is very important when using colorimetrics to read the instructions, but it is even more important with the warfare agent tubes as in some cases they are used differently than regular colorimetric tubes. In most cases the tubes used to detect the warfare agents are detecting the hydrolysis products of the warfare agent, and not the agent itself, so a positive does not mean that the warfare agent is present, only the fact that a hydrolysis agent is present. One major plus for the colorimetric tubes is that they have the fewest false positives when dealing with the warfare agents. Generally if they indicate the presence of warfare agents, they are most likely present. A listing

Figure 11-4 Colorimetric tubes are one of the best devices that could be used for the detection of terrorism agents.

TABLE 11-2

Warfare Agents and Colorimetric Tubes	
Warfare Agent	**Dräger Colorimetric Tube**
Lewisite	Organic arsenic compounds
S-mustard	Thioether
N-mustard	Organic basic nitrogen compounds
Sarin	Phosphoric acid esters
Soman	Phosphoric acid esters
Tabun	Phosphoric acid esters
Cyanogen chloride	Cyanogen chloride
Phosgene	Phosgene
Hydrogen cyanide	Hydrocyanic acid

of the Dräger colorimetric tubes for warfare agents is provided in Table 11-2.

NOTE One of the best methods for detecting warfare agents without using an electronic device is to use colorimetric sampling tubes.

NOTE One major plus for the colorimetric tubes is that they have the fewest false positives when dealing with the warfare agents. Generally, if they indicate the presence of warfare agents, they are most likely present.

Direct Read Instruments

There are two major divisions of direct read instruments: handheld, portable instruments, and those that are prepositioned for special events. The technology with these two divisions are comparable; the only difference is the box the detector is in. Prepositioned equipment is usually radio or hard wire controlled to a central location. Direct read instruments and their method of operation are listed in Table 11-3.

CHEMICAL AGENT METER. There are several chemical agent meters (**CAMs**) on the market (see Figure 11-5), and they do not all use the same technology. In some cases the instruments look identical

TABLE 11-3

Direct Read Instruments		
Device	**Agents Detected**	**Mode of Operation**
APD2000	Nerve, blister, irritants, radiation, toxic industrial chemicals	Emergency response
CAM	Nerve and blister	Emergency response
ICAM	Nerve and blister	Emergency response
ICAD	Nerve and blister	Emergency response
Saw MiniCad	Nerve, blister, and choking	Emergency response
M8A1	Nerve and inhalable aerosols	Prepositioned and emergency response
GID-3	Nerve, blister, and can be programmed for choking and blood agents	Prepositioned and emergency response
M21 RSCALL	Nerve and blister	Prepositioned
M90	Nerve and blister	Prepositioned and emergency response

Figure 11-5 A chemical agent meter (CAM) being used by Marine CBIRF group to detect a potential terrorism agent during a training exercise. Also in use is a MultiRAE gas and PID combination, and the liquid leak is being checked with M-8 paper.

on the outside but the insides are very different. The American and the British CAM uses technology known as **ion mobility spectrometry** (IMS), a proven technology in the science world. The American CAM is made by Environmental Technology Group (ETG) and the British CAM is made by Grasby Dynamics, both companies being subsidiaries of Smith Industries. A nickel 63 radiation source ionizes the sample gas, the ions travel up a pathway, and the instrument is set to read certain ion particles if they move up the pathway. If the mobility of the gas matches one of the algorithms loaded in the CAM, it indicates that the agent is present. The CAM has a quick response time of less than 1 minute, but is susceptible to poisoning due to overexposure. If the instrument is overexposed, it may take a considerable amount of time to clear the instrument. The CAM only reads one agent at a time, and if another agent is in the area you are sampling it does not have an alarm to warn of the other agent's presence. To determine if the other agent is present you have to toggle between the two agents. The CAM indicates the presence of an agent by a bar display on the face of the meter. The bars do not correspond to a specific relationship with parts per million, as one bar indicates a range of material present. Two bars is another range and does not indicate the double of one bar. The military uses the bars as an indication as to when to don PPE.

Some of the problems with the CAM are the fact that it provides false positives for a large number of chemicals. This device was designed and built for the battlefield, not the urban environment where it is now being used. It reacts to aromatics, solvents, and some cleaning solutions. As the CAM uses a nickel 63 source, it used to require the owner to apply for and hold a nuclear device license, a potential problem for some agencies. This issue has been resolved and the company that manufactures the detectors now holds the license for the detector owners. The battery is also a potential problem, as it is a sulfur dioxide battery and is considered hazardous waste, so it can be difficult to dispose of. Users have the option of using regular C cell batteries, a practice that is highly recommended.

The French CAM, which is called the **AP2C,** looks identical to the American and British CAMs but uses flame spectrophotometry as its detection technology. The detection principle behind the AP2C is along the same line as a regular flame ionization detector, as it uses a hydrogen flame to ionize the gas sample. The difference is that the AP2C has algorithms loaded in it that resemble the warfare agents, and if the gas sample matches those algorithms then a reading is provided. It is a good idea to pair up the AP2C with ion mobility technology as they give different false positives. If both technologies report that the unidentified substance is a nerve agent, then it is most likely a nerve agent. If only one meter indicates a nerve agent, then it is probably not a nerve agent, but one of the many materials that causes that particular technology to false alarm.

ADVANCED PORTABLE DETECTOR 2000. The APD2000 (see Figure 11-6) is a CAMlike device in a different style box, with the biggest difference being the software that drives the algorithms. The new software has dramatically improved the false positive issues associated with the CAM, however it still will give false positives for some items, especially cleaning solutions. An invaluable benefit of the APD2000 developed by ETG is that it now has the ability to detect mace and pepper spray, hours after the discharge of the spray and can detect small amounts in the air. Responding to the mall or

Figure 11-6 The APD2000, which detects terrorism agents, mace/pepper spray, and toxic industrial chemicals.

school was always frustrating as experienced responders knew the release was mace or pepper spray but colorimetric tubes were not helpful unless they arrived quickly. This detector has solved a lot of HAZMAT issues when dealing with this kind of event. One other option that is available to responders is to have the ability to detect gamma radiation with the APD2000.

The false positive issue remains even with this much improved detector, as it is the nature of the technology. The ion mobility looks at peaks and length of spikes produced when the sample gas is run through the chamber. These peaks as shown in Figure 11-7 are compared to peaks in the software. If the peaks resemble the peaks in the software, then the detector alarms. The software's algorithms are fairly tight, and only chemicals that closely resemble the desired gases cause an alarm. The problem is that the algorithms for the warfare agents are looking for indications of phosphorus and sulfur molecules in the chemical makeup of the sample gas. If the sample gas has sulfur or phosphorus as part of its makeup then it is likely that if any other component of the molecule has any resemblance to a warfare agent the detector will alarm. The detector alarms when near organophosphate pesticides, which it should as the chemical structures are identical to nerve agents. Other problematic chemicals are cleaning solutions, usually those for carpets and tile floors. They usually contain phosphorus or sulfur and have other components that mimic warfare agents. It is technically feasible to engineer some of these false positives out, but the tighter the algorithms become the more there is a chance that the unit could provide a false negative. A false negative means the detector did not alert to the presence of an agent that it should alert to. We can deal better with a false positive, and if we treat a false positive as a true event, there is less chance of harm than there is for a detector to say that the event is safe when it is not safe. There are too many variables in the types of warfare agent mixtures that a terrorist could come up with, and it is better to have some false positives than to risk missing an off-kilter nerve agent mixture that may not be picked up by a tight algorithm. The biggest problem with false positives comes with the detection of VX. Because VX has a very low vapor pressure, the algorithms are fairly wide to make sure it picks up any hint of VX. The APD has the ability to detect VX separately, which would be recommended as it limits the number of false positives.

The APD2000 can be hardwired or radio connected to a central location, such as a laptop, to report the readings on the detector. Through this kind of use units can be prepositioned around a building or a stadium and remotely monitored. There are plans in the future to have the APD2000 provide detection for a menu of toxic industrial chemicals. The ability to detect toxic industrial chemicals to the parts per billion range is an extreme advantage for this device, as it becomes even more useful for standard HAZMAT response.

IMPROVED CHEMICAL AGENT METER. The only difference between an **improved chemical agent meter (ICAM)** and a CAM is its size and the configuration of the detector. The ICAM produced by ETG was put into a different style box than the CAM to make it easier to repair. The ICAM is also more suited to be prepositioned than the CAM, and that was its most typical use. The IACM can be hardwired or radio connected to a central reporting or tracking station. The ICAMs could be used to ring a stadium prior to an event, all reporting to a central location reporting any readings picked up. The only drawback with this kind of use is that it has the same false positives as the CAM.

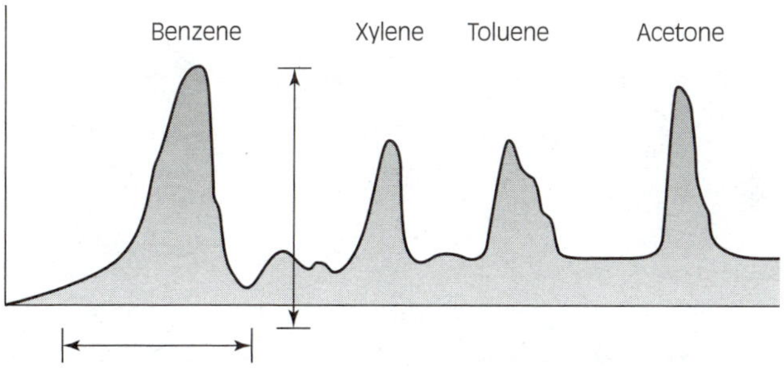

Benzene Xylene Toluene Acetone

Figure 11-7 Ion mobility provides peaks, which are measured for height and width, and compared to a library of other peaks. If the peaks match or are close then the meter alarms.

SURFACE ACOUSTIC WAVE. The device that uses **surface acoustic wave (SAW)** technology is the SAW MiniCAD mk II and its improved version HAZMAT CAD and HAZMAT CAD Plus. Both use a fairly unique detection technology produced by Microsensor Systems. The SAW MiniCad is not IMS, but the SAW sensors produce a similar algorithm, which is checked twice for confirmation. SAW is not a common device as it was not used by the military, but its unique technology makes it an attractive device. The SAW has two piezoelectric crystals that are absorbed into the coatings on the sensor surface. This absorption causes a change in the resonance frequency of the instrument. This resonance outputs an algorithm much like those shown in Figure 11-7, except in this case two sensors are outputting algorithms. The two readings are compared against those stored in the library, and if a close match is found then the unit alarms. The sensors then heat up to remove the sample agent from the sensors. The SAW has many of the features that other civilian detectors have, including alarms and datalogging. The SAW also has the ability to detect both nerve and blister agent at the same time, and does not require toggling between the two agents. The HAZMAT CAD Plus adds the ability to detect some toxic industrial chemicals through the use of electrochemical sensors. The CAM, APD2000, and the ICAM only detect one agent at a time, and the user must switch between modes to see if any other materials are present. The SAW has about the same number of false positives as the APD2000 and is susceptible to the same type of false positive problems, although it is a different set of false positives. The combination of an APD2000, AP2C, and a SAW would make a great set of detection devices. If all three alarmed for the same agent then you could be reasonably certain it was that agent. If one of the three did not alarm, then the material would most likely not be the warfare agent, as they all react to differing interferents.

GID-3 CHEMICAL AGENT DETECTOR. The **GID-3** detector (Figure 11-8) also uses ion mobility spectrometry (IMS) as its detection technology. It is produced by Grasby Dynamics which also makes the British CAM so the backbone to this device will mimic the CAM. GID-3s have improved algorithms so as to reduce the number of false positives, but

Figure 11-8 The GID-3 meter detects nerve and blister agents.

have false positives comparable to the CAM. It is used normally in vehicles, but can also be set up to monitor perimeters in an emergency response mode. It can also be hardwired or radio connected for prepositioned events. Several U.S. military units use the GID-3 as part of their detection capability, so it is possible to see this device in use.

M90 CHEMICAL AGENT SYSTEM. The **M90 chemical agent system** is another device used by U.S. troops and is commonly prepositioned. Made by Environics Oy of Finland, this device also uses ion mobility as its detection technology. It is usually prepositioned but also may be used for emergency response. One unique feature of this device is that as an option you can purchase alerting pagers that can be issued to personnel and when the device is activated the pagers will also alert. This device detects nerve and blister agents, and is subject to the same false positives other military detection devices are troubled by.

Military Test Kits

The most common military test kit used by civilians is the **M256A1 kit** as shown in Figure 11-9. A Canadian kit known as the **C2 kit** is in use, but is not that common in the United States. The C2 is comparable to the US M18A2 kit which is no longer available. Both kits have cumbersome instructions, are difficult to use especially with PPE, and require regular training. Both kits employ wet chemistry as the detection method, and use a color change on a detection strip to indicate the presence of a warfare agent.

The M256A1 kit uses wet chemistry to determine the presence of nerve, blister and blood agents in the air. The kit takes about twenty to twenty-five minutes to go through each of the steps, and requires that specific steps be followed in order to be successful. The test card must be held in certain configurations for varied time periods, and chemical-filled vials are broken open during various times during the test. This kit is the only nonelectronic detection device that detects vapor in the air, which is an advantage. Its relative low cost is a benefit, but its difficulty in use is a severe drawback. Instructions come with the package, and each test kit has a set of instructions. It is best to photocopy the color instructions included with the kit and enlarge them. Have the enlarged instructions laminated and provide one to the entry team and one to someone radioing the instructions to the entry team. Someone needs to have access to a stop watch as several tests must be timed.

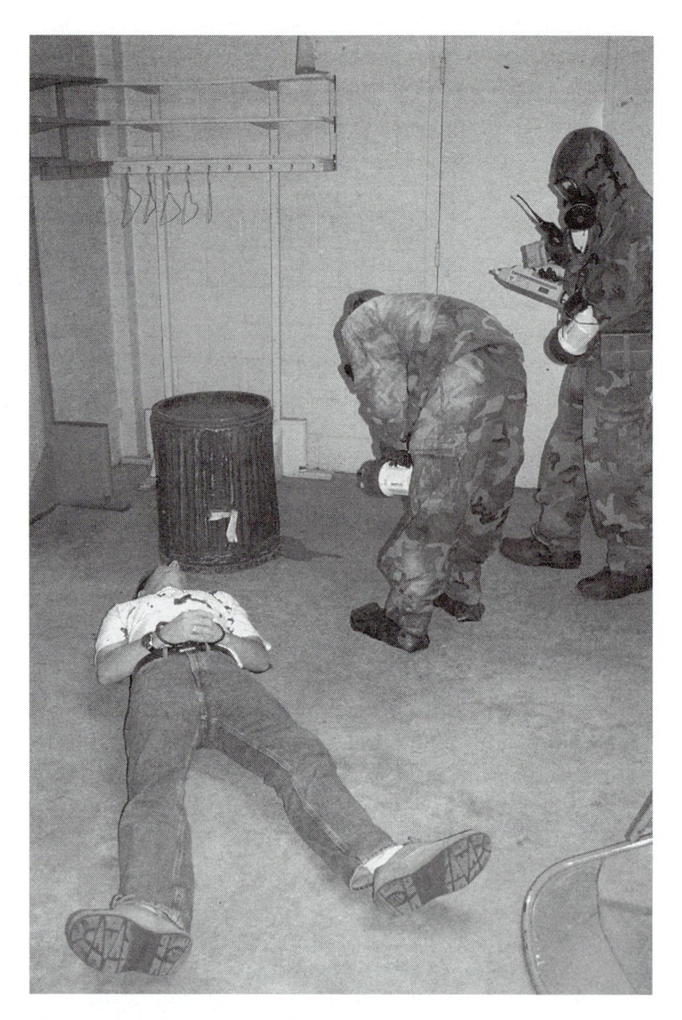

Figure 11-9 The M256A1 kit detects nerve, blister, and choking vapors.

Biological Detection

One of the most common terrorism threats involves the use of biological agents, but this threat has only a minimally effective method of street level detection. There are a number of methods available to responders, from handheld immunoassays known as **smart tickets** to a $200,000 **biological indicating device (BID).** The high end is a very effective method that is used by the military to detect biological agents, but obviously is very costly, both in initial purchase and upkeep. It is relatively easy to use, once one receives some basic training. The smart tickets are relatively inexpensive, but have some drawbacks. To be accurate in the field, extensive training must be provided in sampling and testing techniques. Not only are false positives possible, but it is very easy to have false negatives, in other words the real biological material is present but the detection device did not indicate its presence. Although more recent versions have tried to eliminate some of these problems, there are many concerns when doing this type of sampling. Some street level tests involve electronic sampling using bioluminescence (bacteria) and spectraphotometry (toxins) to indicate the presence of biological agents. One of the major issues is the fact that biological matter exists throughout the community, and the detection devices have difficulty differentiating between naturally occurring harmless materials and a biological warfare agent. After explosives, biological threats are common, and one should be concerned about the ability to detect these materials. Luckily the FBI has established fifty-six laboratories strategically placed around the country for the purpose of detecting biological warfare agents. These labs meet the FBI standards for sampling and identification of biological materials. Within a few minutes these labs can provide an initial idea of the credibility of the threat and a short time later can provide additional confirmation. It may take a while to provide a positive identification of what the material actually is, but they can easily tell you what it is not. To have a sample analyzed at one of these labs, contact your local FBI office and talk with the weapons of mass destruction (WMD) coordinator. In order for the FBI to do testing, one of the first questions that needs to be answered is whether and what kind of crime has been committed. As an example, these FBI labs do not do biological sampling for a sick building if a crime has not been committed.

SAMPLING STRATEGIES FOR TERRORISM AGENTS

Of all the tasks a HAZMAT team may get involved in, terrorism has great potential to be a literal minefield of concerns. There will be great pressure from a number of fronts to identify the potential agent and to assist in the mitigation of the event. There is potential for large numbers of patients, who may or may not be cooperative, and there may be considerable hysteria with the victims and responders. In the United States there are about 3000 actual bombings each year; the next greatest number of terrorism-related events are hoaxes. The emergency response to one of these hoaxes determines whether the terrorist is successful. To be successful the terrorist does not actually need to have Sarin nerve agent, but only needs to have you think he has it. How we respond to this potential threat directly determines who wins, and obviously we need to win this battle. When dealing with a potential terrorism event we need to think as though we are at war, a unique and troubling concept.

Detection and Protection is a concept that was codeveloped with Frank Docimo for a National Fire Academy terrorism program. This concept is that detection is the first priority while wearing an appropriate level of protection. Responders need to wear appropriate levels of protection that take into account the chemical, the amount of chemical, chemical risk category, heat stress, psychological stress, work task, and an overall risk/benefit analysis. In many cases only the first item, the chemical, is considered, which endangers the life of the entry team by generally exceeding the actual type of PPE required to protect the responder. Safe but very rapid identification is necessary. The HAZMAT team must focus on identification, which can solve many of the issues. The HAZMAT team, as best it can, needs to ignore extraneous issues and focus solely on identification. Training of first responders is necessary to educate them in the fact that the HAZMAT team functions differently than they usually do, and first responders may need to pick up additional duties. The key to dealing with a hoax is to determine that the agent supposedly is present is in fact not present. The first responders may be initiating mass decontamination procedures and if there is no toxic or harmful material present, then decon is not necessary. If the stated agent is not present, or the unidentified is not presenting any risk, the situation can be downplayed and the community can return to somewhat normal conditions.

With the number of false positives that can be indicated, the detection of warfare agents is very difficult. A chemist who is familiar with HAZMAT emergency response is an important asset in the determining the identity of the unknown material. For terrorism events, you can consult with the FBI's Hazardous Materials Response Unit (HMRU) through your local FBI HAZMAT or WMD coordinator. The FBI HMRU can link you up with its chemists and responders who can provide recommendations for sampling strategies. Most FBI field offices have HAZMAT trained agents, and the larger FBI field offices have HAZMAT teams who have training and equipment for detecting warfare agents and other hazardous materials.

When doing any chemical sampling, it is important to follow the considerations given in this section. Each response has varied conditions and each situation is unique. The sampling strategies provided are suggested starting points. A chemist or other technical specialist can assist in the sampling strategy, and the conditions present may dictate a change. Most important, when sampling for warfare agents, be careful of false positives as they are common. Use a variety of tools in the identification of an agent. Do not rely on any one test to make an identification. Use biological indicators (the victims), test strips, and other detection devices. If they all indicate a nerve agent, then the material is most likely a nerve agent. If the biological indicators are not suggesting any problems and the M-8 says it is nerve agent, then it is most likely not nerve agent. Even if more than one indicates nerve agent, it is most likely a false positive. If you use two differing technologies, such as an APD2000 and an AP2C, and both indicate nerve agent, then it is most likely nerve agent, as they react to different false positives. A CAM and an APD2000 have some similarities with false positives, so that combination would not be accurate. The basic rules that need to be followed while sampling any unidentified materials are the following:

- Always wear a minimum of respiratory protection.
- Avoid contact with the product.
- Sample for the risk category (fire, corrosive, or toxic) and always use a four-gas instrument, PID, and pH detection.
- A minimum of emergency decon must be available.
- Grab a sample and do the testing away from the hazard area.
- Colorimetric tubes, M-8, M-9, and the M256A1 kit can be used to further characterize the unidentified material.
- Never guess as to the identity of the material until you have all the results.
- Back up street tests with lab tests, follow chain of custody.
- Consult with a chemist.

In some cases the detection devices for warfare agents have a minimum detection limit above the IDLH. By using several detection devices, you are more than adequately protected, and exposure values are conservative. The more accurate limits are provided in the lethal concentrations and incapacitating values. In all cases standard and military detection devices are able to detect those substances at values well below those established for the warfare agents. Table 11-4 provides a listing of the common agents and their various exposure levels, but the more realistic values are the LCt50 and the Ict50. LCt50 is lethal concentration to time for 50 percent of the exposed population, in this case the time usually means three minutes. Ict50 is incapacitating concentration to time for 50 percent of the exposed population. The time in this case is also three minutes. Incapacitated means the people are unable to help themselves.

WARFARE AGENTS SAMPLING STRATEGY

The detection of warfare agents is based on military grade agent, something which is not likely found in the real world. With this in mind, a loose sampling strategy looking for various items should be adopted, such as those found in Chapter 10. In many cases if the terrorists have developed the

TABLE 11-4

Warfare Agents Exposure Values

Agent	PEL (ppm)	IDLH (ppm)	LCt50 (ppm)	Ict50 (ppm)
Phosgene	0.1	2	791	395
Chlorine	1.0	30	6551	620
Sarin	0.000017	0.03	12	8
Tabun	0.000015	0.03	20–60	45
Soman	0.000004	0.008	9	4
VX	0.0000009	0.0018	3	2
Mustard	0.0005	0.0005	231	30
Lewisite	0.00035	0.00035	141–177	<35
Hydrogen Cyanide	10	45	3600	n/a
Cyanogen Chloride	0.2	Unknown	4375	2784

agent themselves, it will have impurities, which should increase the ability of some detection methods, such as colorimetric tubes. A homemade batch of warfare agent may not get picked up by other detection devices though. The Aum Shinrikyo cult in Tokyo, Japan, used a 37 percent sarin mixture with acetonitrile in hopes of using the high vapor pressure of the acetonitrile to disperse the sarin, a flawed theory.

Some of the hydrolysis products for the nerve agents are common substances and can easily be detected using standard detection devices. The amount of hydrolysis products will vary with each agent but should be present in all but extremely pure substances. VX will always have hydrolysis products no matter how pure the substance is, which aid in its identification. Table 11-5 provides a listing of agents and their hydrolysis products. Figure 11-10 provides a sampling strategy when dealing with nerve agents. When dealing with VX it may be best to pipette some of the liquid into a sample jar, and then heat the sample jar using a hot water bath to 150°–160°F, then sample the vapor space with the detection devices. This method should be used with the other materials as well, especially if no readings were obtained during the first round of sampling. The Spilfyter strip is used predominately to look for fluorine, but the petroleum hydrocarbons test, may indicate other contaminants. The iodine chip test is a quick method of classifying unidentified hydrocarbons.

One of the immediate problems with the nerve agents is that their chemical structure is so close organophosphate pesticides that an exact identification is nearly impossible in the field. The bottom line is, whether you have a military nerve agent or a commercial organophosphate pesticide, it does not matter. The treatment of victims, decontamination, mitigation, and PPE is all the same. The exact identification aids in the criminal aspect of the investigation, but that can wait for the lab results.

Blister Agent Detection

It is important for HAZMAT teams to check for blister agents quickly as they have delayed effects, and victims may not be showing any signs or symptoms after an exposure. Fortunately with blister agents, there are not many common chemical equivalents, so when the majority of the detection devices indicate the presence of a blister agent, it is quite possible that the material is a blister agent. The one factor going against the detection of blister agents is the fact that their vapor pressures are extremely low. Much like VX it may be best to heat the sample to see if that improves the off gassing of hydrolysis products. Some common chemicals will cause military detectors to indicate the presence of blister agents. The most common one is oil of wintergreen (methyl salicylate) which is used as a simulant for blister agent. Anything that has this flavoring in it will indicate on a electronic device. Mouthwash, breath mints, gum, and many candies have methyl salicylate as the wintergreen flavoring. In most cases these items can be confirmed by other tests. In most cases if *all* of the street level detection devices show blister agent, there is a pretty good chance the material is blister agent. The sampling strategy for blister agents is found in Figure 11-11.

The hydrolysis product for the blister agents is hydrochloric acid, for which there are colorimetric tubes. Some generic acid sampling tubes will also indicate by a distinctive color change for hydrochloric acid. The PID or FID must be close to the sample in order for it to read, and it can be anticipated that the readings will be low. Generally, if all of the

TABLE 11-5

Hydrolysis Products of Warfare Agents

Agent	Hydrolysis Product
Sarin	Phosphonic acid and hydrofluoric acid. Sarin is also fluorinated, which also may be indicated.
Tabun	Phosphonic acid, hydrofluoric acid, and cyanides
Soman	Phosphonic acid and hydrofluoric acid
VX Agent	Will always have thiol amine. Phosphonic acid may be present in early stages in a pure agent, and will also be indicated by accompanied heat increase of the mixture.
Blister agents	Hydrochloric acid

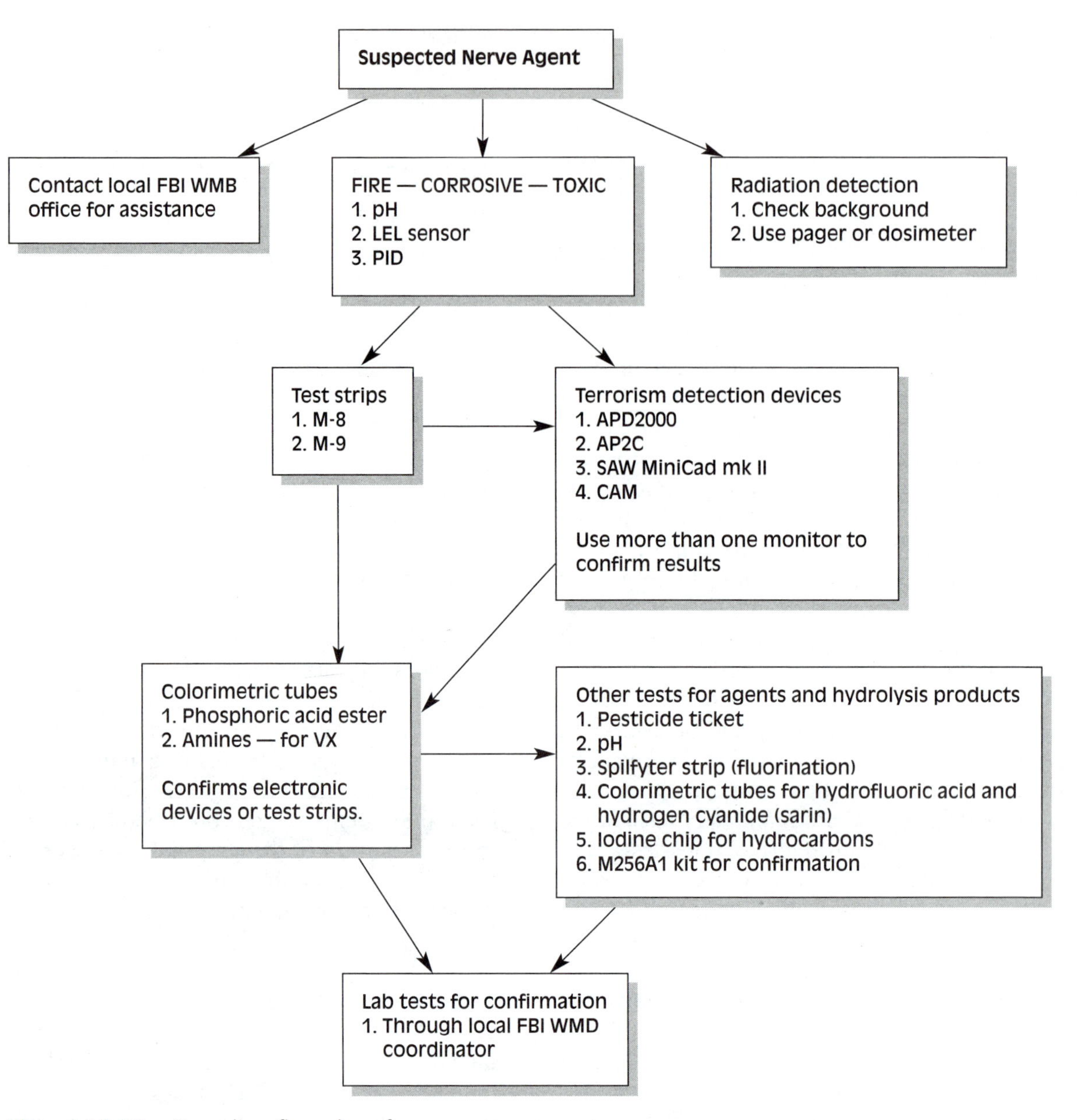

Figure 11-10 Sampling flow chart for nerve agents.

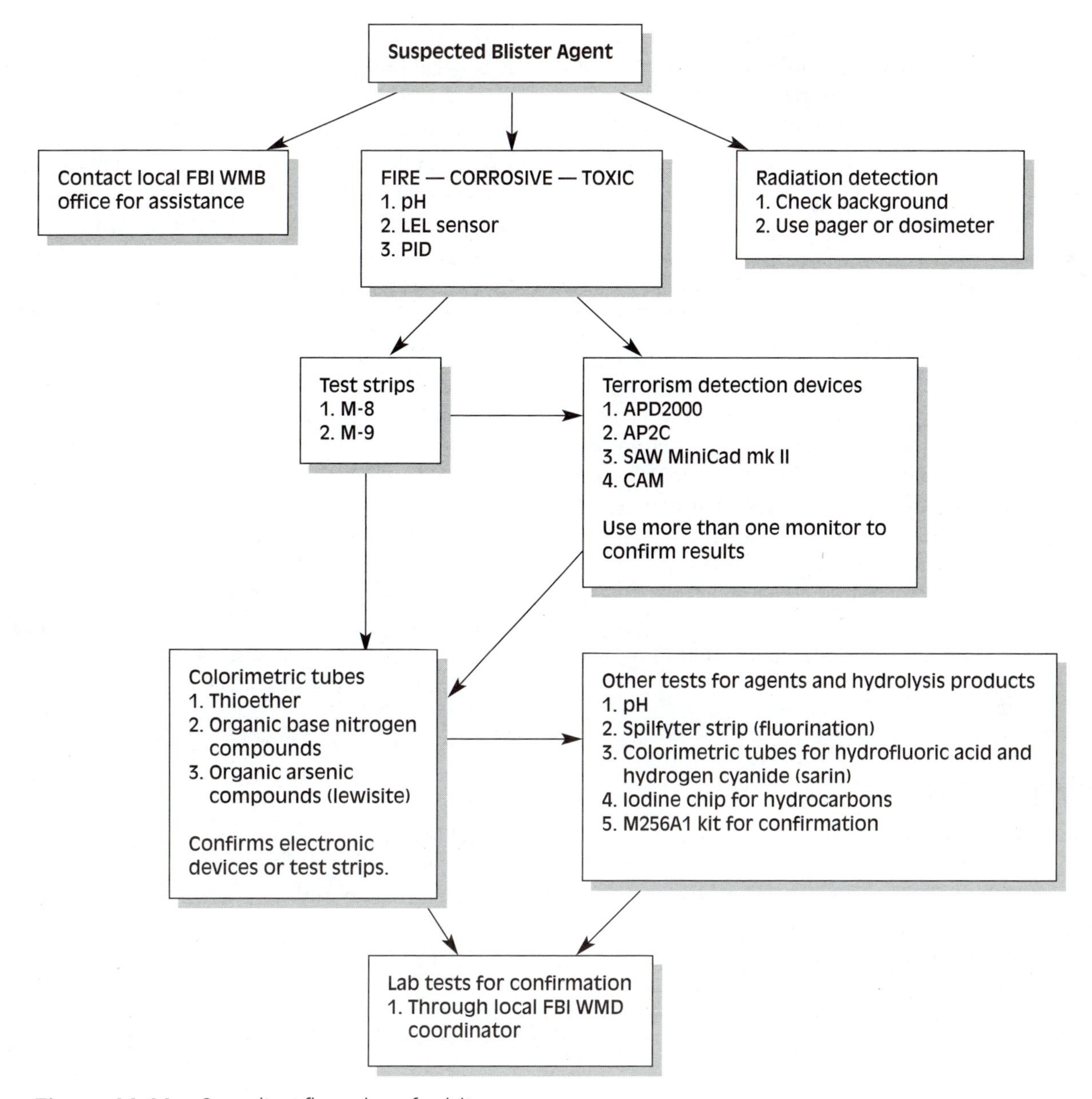

Figure 11-11 Sampling flow chart for blister agents.

military detection equipment indicates a positive for blister agents, then it is probably blister. An agent that provides false positives on all of the devices would be difficult to manufacture, although certainly possible.

SUMMARY

The detection of terrorism agents is more difficult than a standard unidentified material. There are evidence issues, political issues, interagency issues, and potential for a large number of victims counting on your characterization. Request help early and request additional detection devices that can be of assistance. The key to survival in this or any other chemical incident is detection and adequate protection. You can save lives or reduce the hysteria caused by a potential terrorism event if you can quickly characterize an unidentified substance that may be a terrorism agent.

KEY TERMS

AP2C French chemical agent meter (CAM).

APD2000 See *Advanced Portable Detector 2000.*

Advanced Portable Detector 2000 A warfare agent detection device that also monitors for mace, pepper spray, gamma radiation, and some toxic industrial chemicals.

BID See *Biological indicating device.*

Biological indicating device A detection device for biological agents, such as anthrax.

C2 Canadian agent detection kit.

CAM A detection device for nerve and blister agents; there are American, British, and French versions.

GID-3 A detection device for nerve and blister agents.

ICAM See *Improved chemical agent meter.*

Improved chemical agent meter A CAM in a different style box.

IMS See *ion mobility spectrometry.*

Ion mobility See *Ion mobility spectrometry.*

Ion mobility spectrometry A detection technology that measure the travel time of ionized gases down a specific travel path.

M-8 paper A paper detection device like pH paper that detects liquid nerve, blister, and VX agent.

M-9 paper A detection device used for liquid nerve and blister agents; comes in a tape form.

M90 chemical agent system A detection device for chemical warfare agents, was typically used for perimeter monitoring. It has been replaced by the GID-3 monitor.

M256A1 kit A detection kit for nerve, blister, and choking agent vapors.

SAW See *Surface acoustic wave.*

Smart ticket A detection device for some biological agents, such as anthrax.

Surface acoustic wave A sensor technology used in detection devices, used in the SAW and HAZMAT CADS. After the sampled gas passes over the sensor it outputs an algorithm that is checked for a possible match.

VX A nerve agent, more toxic than sarin.

RESPONSE TO SICK BUILDINGS

INTRODUCTION

The HAZMAT response to a **sick building** is unique, but this type of response is well within the capabilities of a well-trained, educated, and equipped HAZMAT team. The key to solving a sick building problem is to first think outside the normal response box. Use ordinary HAZMAT tools in a unique fashion and follow some guidance, but mostly follow your instincts. The cost for a building owner to hire an air quality contractor is extremely expensive. The actual cost to a HAZMAT team is minor, usually less than $100, other than personnel costs. A HAZMAT team that is diversified in its variety of responses is one that will easily survive and, in many cases, thrive in a budgetary crisis.

NOTE The true definition of a sick building is a building that causes the occupants to be affected by an unknown source.

A HAZMAT team can make a difference in a sick building because in many cases the problems found in buildings do not fit the exact definition of a sick building. This chapter focuses on sick building causes that can be detected or identified by a HAZMAT team. The field of indoor air quality is very broad, and this chapter focuses on the most common issues. The true definition of a sick building is a building that causes the occupants to be affected by an unknown source. The health effects can vary from minor irritation of the respiratory system to acute life-threatening illnesses. The source of a sick building is usually the building itself or the contents. The number of sick buildings has increased in the recent times and can be related to a number of causes including building types and an informed public. New buildings are increasingly airtight or sealed buildings, as shown in Figure 12-1. With the increased concern for energy conservation there has been an increase in sick buildings. The main problem is that the buildings do not allow for a good exchange of indoor air with the outside air, allowing a buildup of contaminants. Any small spills that in the past went out the ventilation system or were diluted by outside air coming into the building are now held in the building, causing problems. Problems with sick buildings have come to light because of an informed public and the advent of shock media. One person having a true issue

Figure 12-1 Environmentally tight buildings are one of several reasons for the increased number of sick building responses.

with a chemical exposure in a building can be spread by the media as being a problem to everyone in the building. The media may use environmental issues as a platform to hype a small potential hazard into one that can cause problems to a larger group. There are people who are susceptible to chemical exposures while others are not affected at all. The reaction to a bee sting is a chemical reaction and is one that we can use to discuss the body's response to chemicals. Assemble a group of about fifty people, and within that population group there will probably be one to three people who are allergic to bee stings. They have an immediate chemical reaction to the bee venom while the remaining forty-seven to forty-nine people have no apparent reaction to the bee venom. Just as bee venom affects some people and not others, sick building chemicals have the same type of effect. By exposing a population to a certain amount of a chemical, some will be immediately affected, while others will not show any effect. Isolation of the persons who are affected by a sick building chemical is an important issue as psychogenic illnesses may start to develop. If you keep the people affected by the bee venom in the same group with those who do not have a problem with the venom, you will find that the persons who normally are not affected by bee venom will become symptomatic after time. This sympathetic response is a common problem with sick buildings and is difficult to sort out.

The HAZMAT sick building varies from the previously given definition but is related to a number of causes, some of which are listed in Table 12-1. The HAZMAT sick building is usually related to something the occupants have caused, and is a direct result of some type of chemical activity. The HAZMAT sick building is usually an acute issue as opposed to a chronic issue that a true sick buildings ends up being.

Most of the causes of a HAZMAT sick building are easy to comprehend and a direct factor can be determined. The building can provide any number of potential sources of illness, but it is usually related to the heating, ventilation, and air-conditioning (HVAC) system, which is discussed later in this chapter in the section Ventilation Systems. The furnishings may

ASPECTS OF SICK BUILDING SYNDROME VERSUS BUILDING-RELATED ILLNESS

Sometimes the medical condition that affects the occupants of a sick building is called sick building syndrome (SBS) or building-related illness (BRI). In either case the EPA and the World Health Organization (WHO) have established some criteria to describe both medical conditions. A person who has SBS has symptoms of acute discomfort, with headaches, irritated eyes, nose, throat, dry cough, itchy skin, dizziness, and nausea, difficulty in concentrating, fatigue, and sensitivity to odors. The symptoms cannot be tied to a specific cause or event. The key identifier of someone with SBS is the fact that the symptoms disappear after they exit the building.

Someone with BRI may have a cough, chest discomfort and tightness, fever, chills, and muscle aches. The symptoms have identifiable causes and do not clear after exiting the building. Another term that is becoming associated with sick buildings is multiple chemical sensitivity (MCS) in which it is thought that a person who has worked in a sick building develops severe reactions to many different chemicals. Even exposure to food products, cologne, or chemicals can cause severe reactions to some people. There are multiple issues with each of these described health effects, and it is best to try to avoid labeling a building until all the facts are determined.

TABLE 12-1

HAZMAT Sick Building Causes	
Building	Outside factors
Furnishings	Neighboring facilities
Occupants	Friday 3:00 P.M.
Processes	

also be a cause in a sick building case, as new furniture will off-gas a number of chemicals and, depending on the air exchange rate, may cause a problem with some employees. Carpet is included under the furnishings category, and in many cases is the culprit in sick buildings. Between the adhesive used to lay the carpet and the odor that the carpet itself presents, new carpet can be targeted as a cause of sick buildings. The common gases found in carpets are provided in Table 12-2. It takes about a year for new carpet to off-gas to a point that it no longer becomes an issue. The best time to install new carpet in offices or other commercial buildings is on a weekend, with lots of ventilation, to reduce the irritation to the occupants. The occupants may bring chemicals, usually mace or pepper spray (Figure 12-2) into the building causing problems. Also, the occupants may inadvertently mix some cleaning chemicals which could

Figure 12-2 A small can of pepper spray in a building can create panic and injure large numbers of people.

TABLE 12-2

Volatile Organics Detected in Carpets	
1,1,1 trichloroethene	Ethyl methyl benzenes
1,2 Dichloroethane	Hexanes
1,4 Dioxane	Hexene
4-Phenylcyclohexene	Methylcyclopentane
Acetaldehyde	Methylcyclopentanol
Acetone	Methylene chloride
Benzaldehyde	Octanal
Benzene	Pentanal
Butyl benzyl phthalate	Phenol
Carbon disulfide	Styrene
Chlorobenzene	Tetrachloroethene
Chloroform	Toluene
Dimethylheptanes	Trichloroethene
Ethanol	Trimethylbenzenes
Ethyl acetate	Undecanes
Ethylbenzene	Xylenes

Source: Thad Godish, 1995, *Sick Buildings*, CRC Press, Boca Raton, FL, p. 121.

A NIOSH study in 1989 in which several hundred buildings were studied for sick building issues found the following:

53 percent had inadequate ventilation

17 percent had chemical issues from within the building

11 percent had chemical issues from outside the building

5 percent had biological issues

3 percent had contaminated fabrics in the building

12 percent was from unknown causes

cause problems. Even without mixing some cleaning materials, just their use may cause problems.

The industrial process is another issue that frequently arises with sick buildings. In many cases these problems do not manifest themselves on the production floor, but are found in the office area. The personnel working in the office are not used to the chemicals, as opposed to the personnel who work up close with the chemicals. If a new chemical is introduced to a process, it may cause problems at a facility, as personnel are not used to the odor. If there is a problem with the HVAC system, there is potential for odor migration into the office area causing problems. In a strip store setup there may be a chemical process within the strip of businesses, but an occupancy three doors down is reporting the problem. Several potential causes for this type of problem are the following:

- There may have been a change in the process resulting in more chemicals in the atmosphere.
- The HVAC system may be malfunctioning, allowing migration into other occupancies.
- There may be a common attic allowing migration into other occupancies.
- There may not be good business-to-business wall separation, causing migration.

Outside factors and neighboring facilities can actually be combined, but can be related to a number of issues. The predominant one is chemical activity that is occurring outside the building. Another building may be getting a new roof, and the tar odor can migrate to other locations. Some chemical processes involve a release of odors, which under normal circumstances are not an issue, but occasionally present problems. A weather inversion, which holds things low to the ground, may allow migration of odors into adjacent buildings. A malfunction or change in the HVAC system may draw

The time frame associated with a sick building is a clue as to the possible source. The assignment of a time frame establishes a starting point for the investigation.

Time Aspect	Time Measurement	Common Causes	Usual Source	Assistance Required
Acute	Minutes	Chemical, mace/pepper spray, terrorism	Spill, or other nonroutine release. Cleaning chemicals, construction work. Chemical attack.	Usually none, HAZMAT team can handle.
Acute	Hours	Chemical	Cleaning or construction, such as roofing, painting, carpet installation.	Usually none, HAZMAT team can handle. Low levels may require long-term sampling.
Chronic	Days	Chemical or Allergen	Construction, renovations, or cleaning. HVAC system	Indoor air quality specialist is usually required for a follow-up.
Chronic	Months	Allergen or Biological	HVAC system, building components	Indoor air quality specialist is required.

odors into adjacent buildings causing problems. One other important outside factor involves allergens, which are discussed later in this chapter in the section Allergens.

The Friday at 3:00 P.M. consideration plays an important part in HAZMAT sick buildings. In many areas of the country we can look to the first warm day of spring as a potential response day. Much like someone pulling a fire alarm to get out of work or school, the report of a sick building usually results in the building being evacuated and may result in the employees being sent home. With the news media hyping sick buildings syndromes, these issues have become a big deal. Without an effective response and with an unidentified cause, the building owner or business manager may send the employees home for an early weekend. Labor and management are another issue that fits into this category. Separately ask both groups in the most diplomatic fashion about the status of the labor-management relationship, which may be a factor in why the sick building was reported, and may help identify the cause.

Note: For the remainder of the text we use the term *sick building* in place of HAZMAT sick building.

SICK BUILDING CATEGORIES

The sick building categories include acute, chronic on a short-term basis, and chronic on a long-term basis. The ability to recognize whether the problem is acute or chronic helps identify the problem. An acute sick building problem can be life threatening, and usually does not meet the definition of a true sick building. A HAZMAT team can identify the cause of an acute sick building as it is usually related to a chemical release within the building. Terrorism can cause an acute sick building type, and can be used to panic the occupants of the building.

There are two types of chronic sick buildings: short term and long term. Short-term hazards are usually chemically related, and can also usually be identified by a HAZMAT team. The aspect of short-term time is related to hours and days. The true sick building cause can be difficult to identify, is long term and relates to time in terms of weeks, months, and years. The causes of a long-term type of sick building are usually related to allergens or biological issues. The HAZMAT team may not be able to identify the exact cause but can usually point a building manager in the correct direction to make that identification.

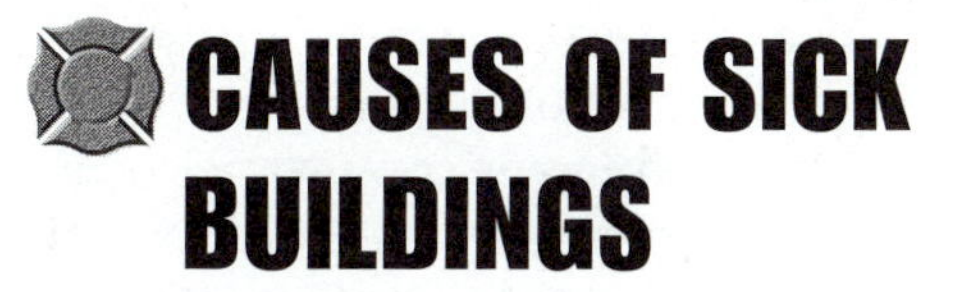

CAUSES OF SICK BUILDINGS

Sick buildings are basically the result of problems with the HVAC system, environmental factors, or contamination.

Ventilation Systems

Identifying the building type is an important consideration in identifying a SB problem. The HVAC system varies with the types of buildings and the type of HVAC system is a key issue with SB identification. The various types of buildings include residential (single- and multiunit), commercial, and high-rise. Although this is a simplification of the many types of buildings, the reason for this small grouping is the HVAC systems. The systems used in these three groupings are different from one another and help identify potential problems. The three systems vary by one basic principle, which is the amount of outside air that the units bring in. In general, residential heating and cooling systems do not bring in outside air. The units heat and cool the inside air and recycle it within the building. With the heating system, units that use some form of flame to heat bring in some outside air to allow combustion to occur. In reality there is some outside air being brought in by all systems, but even at the maximum amount it is not of quantity to be of any concern. The fresh air exchange within a home relies on fresh air coming in from windows and doors. In this age of environmentally sound buildings, this can be problematic if there is limited movement through the doors. There can be the spread of a contaminant throughout a building using a residential HVAC system, but the source of the problem exists within the building.

Commercial HVAC systems are divided into several categories, but for the most part mimic the residential system. Depending on the building size and makeup, the HVAC system is tailored for the building type. In the case of a single occupancy, the likelihood is that there will be only one system, and it will be much like a residential system, just larger. There will be limited fresh air brought in, although a more sophisticated system allows for a maximum amount of 20 percent of fresh air to be brought in although this amount should be considered extreme. Other commercial occupancies, such as a strip mall, generally have multiple HVAC systems, one or more for each occupancy. There is a possibility that odors from one occupancy can be picked up by another tenant's HVAC system, although this would be a rare occurrence as these systems are generally set up to bring in a minimum of fresh air.

High-rise buildings and larger commercial occupancies have systems designed to bring in fresh air to the maximum of 20 percent, but the amount is usually much less than that. These buildings and any other HVAC system do not bring in a lot of fresh air because of cost. There is a cost to heat or cool air, and when the HVAC system is running it is much cheaper to cool the inside air, which should be near 70°F as opposed to cooling the outside air, which may be 90°F. The HVAC standards set by the American Society of Heating, Refrigeration, and Air Conditioning Engineers (ASHRAE) require that outside air be brought in and that a certain amount of air exchange take place. ASHRAE recommends that an air exchange of 20 cfm/person be done in an office, and 15 cfm/person. Areas such as smoking lounges require higher exchanges. These commercial systems vary the amount of outside air coming in by computer control so as to maximize cost savings to the building owner. One downside from the terrorism perspective is that these systems can be overridden by a knowledgeable person to bring in 100 percent outside air. This air would be quickly detected, however, as the air would immediately change, alerting the occupants to a problem in the building. They would not necessarily recognize terrorism, but it would immediately alert them that there was a problem with the HVAC system. Indoor air quality is discussed later in this section, but the key is that people in the building know when there is a problem, and can identify the source.

Understanding how the system is set up is key to identifying the potential source of a SB event. Figure 12-3 provides a diagram of a typical high-rise HVAC system and identifies the various parts of the system. There are several main parts of the system that are standard, building to building, HVAC system to system. These parts include fresh air intake, return air, mixing box, heating/cooling system, and the main fan. From the terrorism perspective there are good places to place an agent for dispersion, and conversely there are places that would not distribute the agent at all. Unless altered, the fresh air intake brings in a minimum amount of air, which is used to make an exchange with the return air from the building. When the temperature is near 70°F, the system will bring in more fresh air as it does not require any heating or cooling. When the temperature varies from that point is when the percentage starts to drop. Even on a day with 70°F temperatures if the humidity is extremely high, the system will not bring in a lot of air, as it would have to dehumidify it. On a high-rise the fresh air intake is located in one of two places: the street level or the roof. When trying to locate the source of an unknown odor or other problem in a building the fresh air intake is a good place to start. One would think that this would be a good place to place a terrorism agent, but luckily this thought process is flawed. First it only brings in a minimum of fresh air, which is then diluted with massive amounts of fresh air, which is then mixed with the return air, which in the smallest percentage would be 80 percent of the air moving in the building. The size of the building needs to be taken into account and the extreme amount of dilution that would occur when a contaminant is placed into that environment. Another factor going against products moving up a fresh air intake is related to physics. The contaminant has to be pulled to the top of the building, which, depending on the physical makeup of the chemical, may present some challenges. Some systems do have a

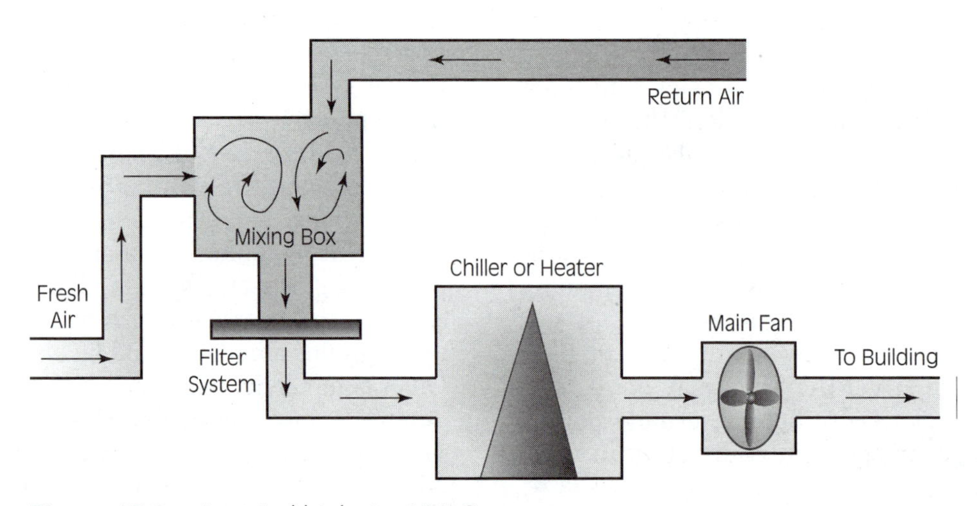

Figure 12-3 A typical high-rise HVAC system.

great amount of pull, but keep in mind the height of the building and the size (and therefore the volume of air) of the air intake. Air intakes on the roof present a fairly easy target for contaminants as the system is pulling the contaminants down into the building, which is easy. We must take into account dilution, and the fact that the most it would be bringing in is 20 percent. For sick buildings there are exhaust vents for various systems, some of which may be venting chemicals, all of which can be drawn into the fresh air intake on a roof. Weather may be a factor in contaminants entering the system, as a weather condition that involves high humidity, fog, or an inversion will keep the contaminants low and hanging around the roof.

The return air system is another location into which chemicals can be drawn and spread into other parts of the building. Placing an agent here is more likely than the fresh air intakes, but some protection mechanisms are still in place. All of the return air is mixed together, which is a tremendous volume of air, which is also mixed with the fresh air. The same complications exist for a return air system, where a contaminant introduced on the first floor has to travel to the top floor all the while it is being mixed with other air, being diluted. The chemical and physical properties have to be such to be conducive to move in that direction. The draw of the return air is not very active and will not "hold" onto contaminants very easily once introduced to the system.

Once the fresh air and the return air are mixed they are then filtered. Any particulate that may have entered the system should be caught by the filters, as long as they are well maintained and installed correctly. During allergy season, an easy fix may be to change the filters as they may have been clogged with the allergens, dust, and/or dirt. If you think a particulate has been introduced to the system, the first place to check is the filters. Most commercial (and residential) filters are designed to trap tiny-sized particles, and will capture almost all contaminants that may have entered the system.

Once filtered, the air then enters the chiller (or heater) where it is cooled or heated. The chiller is a system of pipes that has cold freon moving through it, causing the pipes to be very cold. Part of this process, though, involves that fact that water (condensate) will be running down the chillers, generally into a drip pan. This water is significant as it will trap and hold any contaminants that may have gotten past the filter system. Any chemicals that are in the air will also be caught by this water system and held or altered. Chemicals that are easily broken down or are soluble in water will be caught by this portion of the system. The chillers may be part of the problem as this is a common location for bacteria and other growth to be found. There should be antibacterial tablets in the drip pans to minimize the bacteria buildup. In some cases HVAC personnel have placed HTH™ tablets in the drip pans, which causes large amounts of chlorine to be introduced into the system, presenting a great risk to the occupants of the building. These tablets are made for ten thousand to twenty thousand gallons of water, not the ten gallons that may be found in a drip pan. They make specific tablets for these systems, and they are the only chemicals that should be introduced in the drip pan. If you find bacteria or other growth in the chiller or in the drip pans, you may have found your problem. A good place to look for bacteria is on the discharge side of the chiller area, as this is not easily accessed and may not be cleaned regularly. A good HVAC system does not have any growth in the ventilation shafts or other parts of the duct. In winter, the coils can be 300–400°F, which will alter many chemicals as they enter the system. Even in the heating season there should not be any growth anywhere in the system, and if you find some, you may have solved the problem.

After the air is cooled or heated it is then grabbed by the main circulation fan, which pushes the air to the building. In many systems this is the only fan, and through tapering and shaft sizes the amount of distributed air is modified. Some buildings have helper fans that allow individual floors or work areas to increase or decrease the amount of air flow to that area. Although rare, there have been cases in which these individual areas have been the source of the problem, and they should be examined for contamination. In many cases these helper fans have belt problems, emitting a burning rubber type odor. If the odor is isolated to a specific work area or a floor, then it is the helper fan. If the odor is buildingwide, or on multiple floors, then the problem lies with the main fan. The best source of information about the buildings HVAC system is the mechanical engineer or the maintenance staff. As soon as you arrive at the event, you need to grab these people and hold onto them. They not only can answer questions about the system, but will be knowledgeable about other activities in the building that are likely candidates for the cause.

Environmental Factors

Many of the issues related to response to a sick building are related to the quality of the indoor air. Two of the biggest factors involved in this quality are temperature and humidity level. Just a few degrees off and all of the occupants of the building will feel the effect.

NOTE Two of the biggest factors involved in indoor air quality are the temperature and the humidity level.

Humidity has two extremes, which may be a problem, as too much humidity combined with an elevated temperature will cause many people to feel ill.

The amount of carbon dioxide (CO_2) in the building is a clue as to how efficiently the HVAC system is functioning. Checking the level of CO_2 inside the building as compared to the outside can be an indication of the air exchange. There are no hard-and-fast rules to what will be found in a building, as each building is different. There will always be some difference inside as compared to the outside, with the inside having more CO_2, generally 200–400 ppm more. A difference of more than 400 may indicate a problem. A building that has more than 1,000 ppm may also have a problem. There are conditions that would cause a buildup in a building such as an inversion, but the inside and the outside will have comparable amounts of CO_2. It is important to make sure the CO_2 is coming from an inefficient HVAC system as opposed to a leak of CO_2 within the building. When you have elevated CO_2 in a building, people develop headaches, nausea, and general malaise (tiredness).

NOTE The amount of carbon dioxide (CO_2) in the building is a clue as to how efficiently the HVAC system is functioning.

Contamination

The types of contaminants can vary within a building from allergens, bacteria/viruses, and chemicals. Responders can identify as well as quantify some of these problems while for other problems the only task available to a responder is one of possible characterization. It is sometimes hard to identify a difference between the three large categories, but each has a distinct mechanism of harm and form of identification.

CHEMICALS. Chemicals in a building are the easiest contaminants for responders to identify and quantify, but are also difficult to deal with as the amounts in the air are usually quite low. There are a variety of issues that arise when dealing with chemical problems in a building. Usually the occupants have done something that has caused the problem, so a simple investigation can usually help solve the problem. When the building itself has chemical problems the investigation becomes much more difficult.

The typical scenario for an occupant-caused problem is usually cleaning material or construction. The amount of fresh air being brought in will determine how much of the contaminant will be spread throughout the building. One of the major issues is that in this shock media-educated public is the fact that the public assumes that because they can smell a chemical this smell equates to a direct harm to their body. The longer this believed harm occurs, the more you are unlikely to convince them otherwise. Many chemicals that have potential to cause harm, have odor thresholds below that of the level where they begin to cause harm. Some people who are more sensitive to chemical exposures than others may incite mass illness among the other occupants of the building. It is important to separate out those individuals who are having problems from those who are not having any signs or symptoms. Leaving them together will invariably cause the nonsymptomatic folks to suddenly pick up the signs and symptoms of the other occupants.

One of the issues with painting, resurfacing, or sealing of floors, walls, and ceilings is that the odor is usually pretty irritating, but may be below the detectable limits of most detection devices. The PPB photoionization detector will pick up these odors however and is an essential tool for sick building response. A standard photoionization detector will not pick these levels up unless they are at higher amounts. When cleaning chemicals are involved, the amount of chemicals in the air is usually high enough to get readings on standard detection devices. While conducting an air monitoring survey, it is important to interview the maintenance staff to find out what they were doing and what the location was. In some cases, simple carpet cleaning is enough to irritate some occupants of the building. If bathroom cleaning materials were inadvertently mixed there is usually an off-gassing that occurs that will be irritating to some people. If this has occurred, the cleaning staff is usually not willing to admit to this inadvertent mixing, and through detection devices and good questioning techniques the problem can usually be solved. When looking for cleaning chemical-type problems, it is important to search trash cans and dumpsters as the culprits may have tried to hide their mistakes.

Any type of painting, resurfacing, or cleaning that will output irritating odors should be done when there are no occupants in the building, or when there is an absolute minimum of occupants present. Obviously cost drives this issue, but when the building is closed and the complaints are being investigated there is a tremendous cost to the occupant's business as compared to some overtime for the work crews. Many of these cases become workers' compensation issues and are tied up in court for many years, costing considerable amounts of money.

Other chemical problems that occur in buildings are caused by the building or its furnishings. According to the text *Environmental Sampling for Unknowns* studies have found that new buildings can have up to 30 ppm of volatile organic compounds (VOCs) such as those found in Table 12-3. The predominate chemical usually found in these buildings is formaldehyde, along with several other VOCs. After a year, the level drops to less than 1 ppm for the same building. You should not find levels in excess of 1 ppm for VOCs in office type occupancies or other nonindustrial businesses. Many of these issues also apply when remodeling is done as well. New carpets, wall coverings, and floor tiles all release chemicals that have been known to cause irritation in some persons. A list of these materials is found in Table 12-3. The level of chemicals in the air may not be readily identifiable to us, but the occupants in the building are there for extended periods of time. New furniture, such as desks, chairs, couches, and other office accommodations all release VOCs as well, some of which are chronic irritants. Office equipment such as copiers, printers, or other reproduction equipment also releases irritating chemicals. In some businesses there is an office area with an adjacent production area. Look to the production area for the source of the problem. There may be an area where chemicals are used but that is separate from the area in which the problem is being reported. It is important to ask if there is any new chemical use, or if there has been a change in quantity or procedures.

ALLERGENS. **Allergens** are a common problem, and in some cases may fit within the true definition of a sick building. Although a lot of attention is placed on the chemical part of sick buildings, allergens are the most likely culprit. The growth of fungus or the level of fungi in the building air is highly suspect as a leading cause of sick building syndrome. Unfortunately it is the most difficult to detect, and a specialist has to be brought in. In some cases the allergen issue is a transient one and is related to an unusual spike in allergens in the air.

A dirty building that undertakes a cleaning project may also put an unusual amount of allergens in the air. Two big areas that are usually problematic are carpet removal and ceiling tile removal. These projects should be done after hours when no occupants are in the building.

If the HVAC system is dirty and dusty, then it is spreading allergens throughout the building. The filters within the HVAC system are a clue to the cleanliness of the system, and if they are clogged or filled with dust and dirt, they are compromised and allergens are being spread. Some duct work has a fiberglass lining, which if wet, or has been wet in the past may be the host to mold or mildew, which are allergens. If you find evidence of these items, then they can be a suspected cause. There have been no studies linking surface mold and mildew with illness. The presence of mold and mildew indicates a problem, but can not be directly tied to occupant illness. Mold and mildew cause an odor, which can be irritating over time, and that causes health problems with some people. Mold and mildew anywhere have

TABLE 12-3

Common Building Chemical Contaminants	
1,1,1 trichloroethylene	Carbon tetrachloride
Acetone	Methyl ethyl ketone
Benzene	Toluene
Cellosolves	Aldehydes
Cyclohexane	Deacon
Freon 113	Tetrachloroethylene
Methyl isobutyl ketone	Tetrahydrofuran
Pentane	Hexane
Xylene	Limonene
Amines	Aromatic hydrocarbons
Formaldehyde	

the potential to cause health problems throughout smaller buildings. In larger buildings there is still potential to cause problems, but it will likely not be buildingwide. Mold and mildew will cause buildingwide problems if they are in the HVAC system. There should not be any mold, mildew, or other growth anywhere in the HVAC system. Mushrooms growing in the system are a big clue that there is a problem. The cleaning of the HVAC system is another area that may cause problems as the dirt and debris will be spread throughout the building. This cleaning should be done after hours when no one can be affected by it. In some cases the cleaning involves the use of chemicals, which themselves may cause irritation to some people.

SAFETY The presence of mold and mildew indicates a cleanliness problem and cannot be directly tied to occupant illness.

One good question to ask is if there are roof leaks or other water problems, as they are sources of mold and mildew. If there has been a roof leak that has reached a suspended ceiling, then mold and mildew may be growing on the top side of the ceiling tiles. Even if you cannot see a stain on the bottom side of the ceiling tile, it is a good idea to look at a representative number of ceiling tiles to check for contamination. Check the building for other water leaks, such as in basements, as they are another common area for mold and mildew.

BACTERIA/VIRUSES. Viruses generally spread through a building by human contact and normal spread of colds and other viral illnesses. The most likely routes of spread are related to poor housecleaning, contaminated food preparation areas, bathrooms, telephones, and close proximity to the sick people. In rare cases, these viruses can be spread throughout the building by the HVAC system. A common problem within buildings that occurs several times a year is attributed to a bacteria known as *Legionella pneumophila,* commonly known as Legionnaires' disease. Each year there are a number of incidents of Legionnaires' disease that can cause fatalities. Legionnaires' became famous when it was found to have caused approximately 29 deaths and affected 189 people at a American Legion convention at a Philadelphia hotel in 1976. The number of cases varies each year but there have been up to 100,000 outbreaks of Legionnaires'. There are thirty-four known species and fifty subgroups of Legionnaires'. *Legionella pneumophila* thrives in stagnant water. This water can come from a variety of sources such as aerosolized water in misters, hot water heaters, showers, hot tubs, and humidifiers. Any place water stands for a period of time is subject to the growing of *Legionella.* There have been cases in which the sources were grocery store fruit and vegetable misters, fountains, and a hot tub display in a hardware warehouse. The hot tub display case presents an interesting view of how Legionnaires' can be spread. A number of persons became ill in a suburban community, and it was determined that they had all contracted Legionnaires'. There was no common thread with the affected persons, which normally points to a source. After more than thirty days, it was finally determined that all of the affected people had been to a hardware store, which had a hot tub display. The particular tub in question had been sold, and luckily had not yet been installed in the purchaser's home, but was sitting in his garage. Testing showed that the filter system contained Legionnaires', and through contact with the water each of the affected persons contracted Legionnaires'. In 1999, two people died and several more become ill from Legionnaires' disease in a Baltimore manufacturing facility. It was determined that a production line rinsing system contained Legionnaires'.

SAFETY Each year there are a number of incidents of Legionnaires' disease that can cause fatalities.

There are two other biological concerns, both airborne pathogenic bacteria. The most common one is *Aspergillus,* which causes aspergillosis, and the other is *Histoplasma capsulatum,* which causes histoplasmosis. Hospitals are at risk for aspergillus, which results from a fungus growth, of which there are a number of forms. Although aspergillus can be found in many buildings, patients in a hospital are in a weakened condition and are more likely to be affected. The histoplasmosis is of concern for building occupants and firefighters. This bacteria grows in bird droppings, from pigeons or chickens. After a sufficient time the bird droppings dry and become dustlike. When this dust and dirt is disturbed and becomes airborne, persons without respiratory protection are at risk for histoplasmosis. This problem appears in both rural areas and inner cities.

NOTE Any place water stands for a period of time is subject to the growing of *Legionella.*

ACUTE SICK BUILDINGS

There are four basic causes of sick buildings. The four are related to cleaning chemicals, mace or pepper spray, HVAC systems, and building chemicals. Most of the incidents will fit into one or more of these causes. The flow chart provided in Appendix A provides a breakdown of how to proceed for these types of events. By following the chart and using the checklist provided in Appendix B you can solve the majority of the sick building issues. One item to always remember is that even though you may not be able to identify a specific cause, you will be able to identify what is not the cause.

The cleaning chemicals cause focuses on chlorine, ammonia, and acid gases, as they are commonly used or released when cleaning chemicals are the source of the problem. Obviously the storage areas are the focus of the investigation, as is any area that was being cleaned or had been cleaned. The reported odor by the occupants can help guide this type of investigation. If they report a petroleum type odor, then colorimetric tubes for hydrocarbons should be used as well. The readings on a PID are crucial as well when these types of odors are reported.

Mace and pepper spray is a difficult issue to solve as the chemicals are difficult to detect unless you have an APD2000 gas detection device. You can identify mace and pepper spray using colorimetric tubes, but you must be in the building quickly, the building must have been tightly shut, and your response time must be quick. Mace can be detected (the hydrolysis product) by using the chloroformates (Dräger) tube and pepper spray can be detected with the olefines (Dräger) colorimetric tube. Mace is easier to detect than pepper spray, and the pump strokes must be increased to make the tubes more sensitive. The APD2000 can detect small amounts of mace or pepper spray hours after they have been sprayed, even outdoors. One way to determine the use of mace or pepper spray is to do lab tests on clothes or carpet. Find the most affected persons and bag their clothes in a tightly sealed bag. Send the clothes to a lab and have them run tests on the clothes looking for the breakdown products of mace or pepper spray.

If the HVAC system is the cause, we can point the building owner in the right direction, but we cannot identify the specific cause. Temperature and humidity play a major factor in the comfort of the people in the building. Just by asking a few questions and then determining what the actual temperature and humidity are can sometimes solve a small problem. If the problem is the air exchange, then the problem is more difficult to solve. The gas that is present to check air exchange is CO_2, which should be slightly elevated inside a building as compared to the outside air. However, there should not be a large disparity between the inside and outside. More than a one-third difference in the values indicates a potential problem. If more than 1000 ppm CO_2 is found inside a building, then the levels are high enough to cause some irritation with some people. Most buildings have CO_2 levels of 400–700 ppm and have a corresponding level outside of 200–500 ppm. The higher the amount inside means that there is not enough air exchange with the outside air. This issue is economic in some cases, as the more fresh air that is brought in requires additional heating and cooling. Just make sure the CO_2 levels are from buildup in the building not a leaking CO_2 cylinder.

Other HVAC issues relate to the cleanliness of the system, which may indicate an allergen or biological hazard. It is outside the realm of a HAZMAT team to identify the exact type of allergen or biological that may be present, but we can point the building owner in the right direction. By identifying dirty filters, mixing box, chiller/heater, or duct work, a potential problem can be investigated. There should not be any dirt, dust, or growth within an HVAC system. If you find any, then refer the building owner to a HVAC cleaning specialist or indoor air quality contractor.

Related to the cleanliness of the HVAC system is the cleanliness of the building. Any accumulation of dirt or dust in the building, particularly in areas that may be picked up by the HVAC system are always suspect for problems. Any moisture, mold, or mildew on the walls, ceiling, or floor is also an indication that there is moisture in the building.

The chemical issue is the easiest to determine, as responders' detection devices usually pick up a chemical. As detection devices such as the RAE PPB photoionization detector and the APD2000 have detection limits extremely low and in some cases below the odor threshold, the use of these devices can help solve a sick building mystery. Most chemical issues in a building are related to a human factor; someone in the building has done something to cause the problem.

SPECIFIC RESPONSE ACTIONS

Once you have evacuated the building, shut down the HVAC system, and ensured that all the doors and windows are closed, you can start the investigation. Start interviewing the most injured or affected people as they were most likely the closest person to the problem. While these initial interviews are taking place, the air monitors should be warming up, bump tested, and readied. The maintenance person, and/or manager should be interviewed as well, and told to stand by at the command post or HAZMAT unit. If you learn of chemicals stored in the building or during the interviewing you discover that a chemical hazard may be the cause, determine the most effective detection method for those chemicals. Have the air monitoring crews start at the areas where the most affected people were located. If many people are affected or the problem seems to be buildingwide or floorwide, then start your investigation at the HVAC unit, paying attention to the fresh air intakes and the return air duct. When dealing with a sick building, you should keep terrorism in the back of your mind. The probability is low, but many terrorism scenarios can be initiated through a sick building call. A response team that is effective in identifying sick building causes is a step ahead for terrorist attacks on a building.

If you can identify the cause of the sick building, then follow the appropriate strategy first, and follow up as needed with the other strategies. In the spring and summer make freon a high priority for sampling, and in winter make CO a high priority. Always check for these gases out of season, but move them to a lower priority in the off-season. Concentrate on the immediately dangerous gases first, then work to the chronic hazards. When you have run through a sampling scenario, make the colorimetric tubes more sensitive and do some more sampling. Have air monitoring crews meet with the personnel conducting interviews to compare notes. Have a crew member focus on the SB checklist to cover the other areas not covered in the first interviews. It is important to interview someone from management and someone from the labor group. The more people from each side the better, and make sure the interviews are conducted away from other people. It is important to document these interviews as well as document your air monitoring procedures. The air monitoring use report is a good worksheet to record any readings and track the process. The worksheet has the sick building chemicals listed in the colorimetric section to make this sampling easier. This worksheet can then become part of the final report and documentation.

The air monitoring strategy focuses on two areas: the RBR profile to identify serious health risks, and a focused strategy looking for specific chemical or biological hazards.

The focused strategy is divided into four areas; mace/pepper spray, cleaning chemicals, biological, and chemical. If you cannot identify which of these four areas to start with, follow them in order as provided in the worksheet provided in Appendix B. No matter the situation, always take in pH paper, a three- or four-gas instrument, and a PID. The PID is most important to pick up tiny amounts of chemicals in the air and should be monitored most of the time. It is not likely that the LEL sensor will pick anything up in a sick building. The oxygen is also not likely to drop, but the CO is important to watch especially during the initial entry. Any time you change occupancies, floor, or section all of the monitors should be observed for changes. The initial monitoring should be done with the HVAC system shut down. After sampling the building or if you suspect that the HVAC is the source of the problem, have air monitors in place and then turn the system back on. Pay particular attention to the discharge vents and the return air vents. The sampling priorities for the focused strategies are outlined in Table 12-4.

When all the sampling is done, you have covered most hazards usually found in a sick building. If you did not find anything with the suggested strategies, you may want to sample with other available colorimetric tubes. If you sample and did not find anything, that does not mean that it was a false call. It means there are two possibilities: Either you did not have a detection device for the material present, or the level of contaminant is below the minimum detection limit for your detection devices. When you are done and did not find anything, you can be assured nothing present is an acute risk to the occupants or nothing present is immediately dangerous to their life or health. It does not mean that there is not a chronic or long-term hazard present. There are times when you know a material is present, but you are unable to detect it. In these cases, refer the building manager to an indoor air quality specialist. Some HAZMAT teams and some health departments can do long-term sampling. The air in these types of buildings needs to be sampled for a minimum of 8 hours, if not 40 hours, and the gas samples run through a gas chromatograph. If you suspect a biological or allergens, then specific testing needs to be done for these types of materials.

When referring it is recommended that you know something about the companies that you are referring. Many HAZMAT teams maintain lists of hazardous waste contractors that they provide to a spiller, so that a spill can be cleaned up. In most cases the HAZMAT team knows the capabilities of the companies on this list, and in unusual situations can make a recommendation as to which company may be best suited for a particular spill. This type of familiarity is necessary for indoor air quality companies.

When conducting the investigation, it is best to keep all of the employees near the building so that

TABLE 12-4

Focused Sampling Strategies

Mace/ Pepper Spray	Cleaning Chemicals	Unidentified Chemical	Allergens or Biologicals
Evacuate building.	Evacuate building.	Evacuate building.	Evacuate building.
Shut off HVAC system, shut doors, windows.	Shut off HVAC system, shut doors, windows.	Shut off HVAC system, shut doors, windows.	Shut off HVAC system, shut doors, windows.
Start interviews.	Start interviews.	Start interviews.	Start interviews.
Identify source location. Don appropriate PPE.	Identify source location. Don appropriate PPE.	Identify source location. Don appropriate PPE.	Identify source location. Don appropriate PPE.
Follow RBR sampling procedures: pH, PID, ¾ gas, PPB PID, and FID.	Follow RBR sampling procedures: pH, PID, ¾ gas, PPB PID, and FID.	Follow RBR sampling procedures: pH, PID, ¾ gas, PPB PID, APD2000 (irritant mode), and FID.	Follow RBR sampling procedures: pH, PID, ¾ gas, PPB PID, APD2000 (irritant mode), and FID.
Use APD2000 in irritant mode.	Colorimetric tubes: Polytest Chlorine Ammonia Formic acid Hydrochloric acid Ethyl acetate Benzene Acetone Alcohol Note: If no readings are observed, increase sensitivity.	Colorimetric tubes: Polytest Halogenated hydrocarbons Ethyl acetate Benzene Acetone Alcohol Phosgene Toluene Xylene Mercaptan Note: If no readings are observed, increase sensitivity.	Rule out chemical hazards. Use colorimetric tubes for: Polytest Halogenated hydrocarbons Ethyl acetate Formic acid Ammonia Note: If no readings are observed, increase sensitivity.
Colorimetric tubes: Polytest Chloroformates Olefines Note: If no readings are observed, increase sensitivity	Check HVAC system effectiveness with CO_2 and ozone colorimetric tubes.	Recommend further testing by indoor air quality specialists.	Visually examine the building for clues of dirt, dust, mold, mildew, or other moisture-related problems.
Check HVAC system effectiveness with CO_2 and ozone colorimetric tubes.	Recommend further testing by indoor air quality specialists.		Check HVAC system effectiveness with CO_2 and ozone colorimetric tubes.
Recommend further testing by indoor air quality specialists.			Recommend further testing by indoor air quality specialists.

you can ask questions. In some cases there is a desire to send them home, but it is best to keep them around. Every so often have a representative brief them as to what is occurring and the time frame. If you did not find anything, do not tell the employees they are imagining things or that nothing is present. The standard line is "We tested for the following items . . . and we did not find any detectable levels. Based on these results we did not find any acute or immediately life-threatening levels. We did not nor could we rule out long-term chronic issues as that is beyond our capabilities." At this point you should inform the employees of the mitigation plan. If you did not find anything, and you suspect that there really is nothing present (the Friday at 3 P.M. type of incident), then one option is to send the employees home for the day and have the HVAC system run with full 100 percent outside air for a period of time, such as overnight or over a weekend. This does have risks, as the employees may desire yet another day off and will dial 911 to make that happen. The flushing of the building gives the impression that you think something is there, but you just cannot identify it. Go with your instincts. Document what you did and why you did it. Be prepared to defend your actions in court one to five years later, so document very well. The reports, flow charts, and worksheets provided in the appendixes will assist with that effort.

SUMMARY

The response to a sick building can be very frustrating, but at the same time it can be very rewarding, especially when you successfully identify a cause. When you solve a mystery, the employees are happy, the building owner is happy, and the HAZMAT team looks good to the community. The actual cost is minimal, other than personnel which should not be considered anyway as the customer is already paying that bill. The community outreach benefits not only the HAZMAT team, but the citizens as well. The HAZMAT team gets some great practice in air monitoring and the use of colorimetric sampling, with a minimum of stress. There is some initial work, in the form of research that needs to be done so that you can become familiar with the indoor air quality specialists in your area. You may be able use them in other situations, as many have other expertise or specialized detection equipment that many HAZMAT teams do not.

KEY TERMS

Allergens Substances that cause adverse reactions, such as sneezing, in sensitive individuals.

Building-related illness Illness related to a building, the symptoms of which have identifiable causes but do not stop after the person leaves the building.

Multiple chemical sensitivity (MCS) A condition of severe reactions that is thought to result from exposure to various chemicals and other substances.

Sick building A building that is suspected of causing irritation or health problems for the occupants.

Sick building syndrome A temporary condition affecting some building occupants with symptoms of acute discomfor that stop when the person leaves the building.

REFERENCES

Godish, Thad. 1995. *Sick Buildings Definition, Diagnosis, and Mitigation.* CRC Press, Boca Raton, FL.

Hess, Kathleen. 1996. *Environmental Sampling for Unknowns.* CRC Press, Boca Raton, FL.

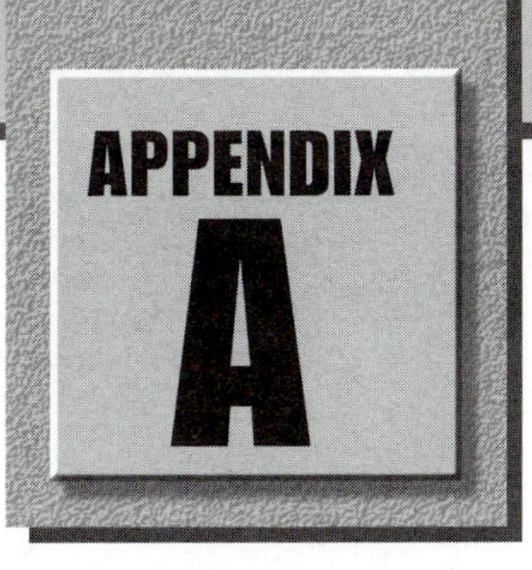

AIR MONITORING USE REPORT

Date	Incident #	Page ________ of ________

Background reading	O_2	LEL	H_2S	CO	MOS	PID

Bump test?	Calibrated?	Operator(s)

Readings	O_2	LEL	CO	H_2S	PID	MOS	Location and time
Initial							
Initial							
Initial							
Initial							
Follow-up							
Follow-up							
Follow-up							
Follow-up							
Follow-up							
Follow-up							
Follow-up							
Follow-up							
Final							
Final							
Final							
Final							

Dräger Air Monitor Report

Tube Name	Sick No.	Number of Tubes Used	Initial Time	Highest Reading	Location	Final Reading and Time
Acetone 100/a	18					
Acetone 100/b						
Air Current						
Alcohol 100/a	19					
Ammonia .25/a	4					
Ammonia 2/a						
Ammonia 5/a						
Benzene 0.5/a						
Benzene 5/b						
Carbon Dioxide O, 1%/a	8					
Carbon Monoxide 10/b						
Chlorine .2/a	5					
Chlorine 0.3/b						
Chloroformates 0.2/b	2					
Cyanogen Chloride 0.25/a						
Diethyl Ether 100/a						
Ethyl Acetate 200/a	12					
Ethylene 50/a						
Ethylene Oxide 1/a	9					
Ethylene Oxide 25/a						
Formaldehyde 0.2/a						
Formaldehyde Activated	10					
Formic Acid	6					
Halogenated Hydrocarbons 100/a	11					
Hydrazine 0.25/a						
Hydrocarbons 0.1%b						
Hydrochloric Acid 1/a	7					
Hydrochloric Acid 50/a						

Tube Name	Sick No.	Number of Tubes Used	Initial Time	Highest Reading	Location	Final Reading and Time
Hydrocyanic Acid 2/a						
Hydrogen Phosphide						
Hydrogen Sulfide 5/b						
Mercaptan 0.5/a	17					
Methyl Bromide 5/b	14					
Nitrous Fumes 0.5/a						
Olefines 0.05%/a	3					
Oxygen						
Petrol Hydrocarbons 10/a	13					
Phosgene 0.02/a	15					
Polytest	1					
Simultaneous Test, acid gases, hydrocyanic acid, CO_2, basic gases, nitrous fumes						
Sim. Test, SO_2, CL_2, H_2S, CO_2, Phosgene						
Styrene 10/a						
Sulfur Dioxide 0.1/a	16					
Sulfur Dioxide 20/a						
Sulfuric Acid 1/a						
Toluene 5/b						
Toluene Diisocyanate 0.02/a						
Trichloroethylene 10/a						
Triethylamine 5/a						
Vinyl Chloride 0.5/b						
Xylene 10/a						

COMMENTS:

COLORIMETRIC SAMPLING STRATEGY FOR SICK BUILDINGS

1. Read the individual tube instruction sheets.
2. Use in conjunction with pH paper (wet and dry), four-gas monitor, and PID.
3. Note initial tube color; use unused tube to compare results.
4. A positive polytest indicates a contaminant is present. If all other tubes are negative, perform further testing.
5. Use Polytest 1st, then if mace/pepper spray is indicated, use tubes 2, 3, and 10.
6. For suspected cleaning agents, use tubes 1, 4–7.
7. Increase pump strokes to increase sensitivity. *** THIS IS IMPORTANT!!!
8. When in doubt, use all available tubes.
9. Record all readings, noting time and specific location. Indicate these on report.

Tube #	Tube Name	Also Reacts To
1	Polytest	Organics and some inorganics
2	Chloroformates	
3	Olefine	Propylene Butylene
4	Ammonia	Amines Hydrazine
5	Chlorine	Bromine Chloride dioxide Nitrogen dioxide
6	Formic acid	Organic acid Inorganic acids
7	Hydrochloric acid	Chlorine Nitrogen dioxide
8	Carbon dioxide	
9	Ethylene oxide 1/a	Formaldehyde Styrene Vinyl acetate Acetaldehyde
10	Formaldehyde	Styrene Vinyl acetate Acetaldehyde Acrolein Diesel fuel Fufuryl alcohol

Tube #	Tube Name	Also Reacts To
11	Halogenated hydrocarbons 100/a	1,1,2 trichlorotrifluorethane (R-113) 1,2-dichlortetrafluorethane (R-114) Trichlorfluormethane (R-11) Monochlorodifluoromethane (R-22) Volatile halogenated hydrocarbons Perchloroethylene
12	Ethyl acetate	Esters of acetic acid Alcohols Ketones Benzene Toluene Petroleum hydrocarbons
13	Petroleum hydrocarbons	n-octane n-hexane n-heptane iso-octane n-nonane Perchloroethylene Carbon monoxide
14	Methyl bromide	Chlorinated hydrocarbons
15	Phosgene	
16	Sulfur dioxide	
17	Mercaptan	Propyl mercaptan n-butyl mercaptan Hydrogen sulfide
18	Acetone	Aldehydes Ketones Ammonia
19	Alcohol	Aldehydes Ether Ketones Esters

This list does not include all the possible items that may be found in a sick building. Each building is different and many other considerations must be taken. The order of sampling provided is for an office building. If any chemicals are present above "normal" office quantities, it is advised to start with those materials. When sampling for CO_2 be sure to record a background reading outside, as CO_2 is always present in the air. It has been implicated on many occasions for causing problems in various types of buildings, including offices and retail areas, but only in high concentrations. Over 1,000 ppm CO_2 indicates a potential problem. More than a 300–400 ppm difference in outside/inside levels may indicate a problem. When sampling refer to the *NIOSH Pocket Guide* for PEL/TLVs and IDLHs for assistance in determining unsafe levels.

If you are unable to locate any hits during your sampling, the next course of action is to increase the sensitivity of the tubes used. By doubling and tripling the pump strokes, you increase the sensitivity of the tube. The next option is to bring in an environmental air quality specialist. They can sample using gas chromatographs, and other long-term sampling media. Sick buildings may result from a large number of items, and the duties of the emergency responders is to determine if there is anything immediately dangerous to life and health in the building. The air quality investigator can determine if there is a chronic long-term problem with the air quality.

SUGGESTED READINGS AND BIBLIOGRAPHY

Bevelacqua, Armando, and Richard Stilp. 1998. *Terrorism Handbook for Operational Responders*. Delmar, a division of Thomson Learning, Albany, NY.

Emergency Response to Terrorism Tactical Considerations: HazMat. 2000. National Fire Academy.

Fire, Frank L. 1986. *Common Sense Approach to Hazardous Materials*. Fire Engineering, New York, NY.

Godish, Thad. 1995. *Sick Buildings Definition, Diagnosis, and Mitigation*. CRC Press, Boca Raton, FL.

Hawley, Chris. 2000. *Hazardous Materials Response and Operations*. Delmar, a division of Thomson Learning, Albany NY.

Hawley, Chris. 2001. *Hazardous Materials Incidents*. Delmar, a division of Thomson Learning, Albany NY.

Hess, Kathleen. 1996. *Environmental Sampling for Unknowns*. CRC Press, Boca Raton, FL.

Lesak, David. 1998. *Hazardous Materials Strategies and Tactics*. Prentice Hall, New Jersey.

Lewis, Sr. Richard J. (editor). 1994. *Rapid Guide to Hazardous Chemicals in the Workplace* 3rd Edition. Van Nostrand Reinhold, New York NY.

Maslansky, Carol J., and Steven P. Maslansky. 1993. *Air Monitoring Instrumentation*. Van Nostrand Reinhold, New York, NY.

McGuire, Steven A., and Carol A. Peabody. 1982. *Working Safely in Gamma Radiation*. Nuclear Regulatory Commission, Washington, DC.

Measurements and Conversions. 1994. Running Press Gem, Running Press Publishers, Philadelphia, PA.

Medical Management of Chemical Casualties. 1995. Chemical Casualty Care Office, Medical Research Institute of Chemical Defense. Aberdeen Proving Grounds, MD.

National Fire Protection Association. 1994. *Fire Protection Guide to Hazardous Materials 11th edition*. National Fire Protection Association, Quincy, MA.

National Institute for Occupational Safety and Health and the Centers for Disease Control. 1999. *NIOSH Pocket Guide to Chemical Hazards*, Washington, DC.

Noll, Gregory, Michael Hildebrand, and James Yvorra. 1995. *Hazardous Materials Managing the Incident*. Fire Protection Publications, Oklahoma University.

Patnaik, Pradyot. 1992. *A Comprehensive Guide to the Hazardous Properties of Chemical Substances*. Van Nostrand Reinhold, New York, NY.

Schnepp, Rob and Paul Gantt. 1999. *Hazardous Materials: Regulations, Response & Site Operations*. Delmar, a division of Thomson Learning, Albany NY.

Smeby L. Charles (editor). 1997. *Hazardous Materials Response Handbook 3rd edition*. National Fire Protection Association, Quincy, MA.

Stilp, Richard, and Armando Bevelacqua. 1997. *Emergency Medical Response to Hazardous Materials Incidents*. Delmar, a division of Thomson Learning, Albany, NY.

Turkington, Robert. 1995. *HazCat Abridged Manual for Field Use*. HazTech Systems Inc., San Francisco, CA.

Turkington, Robert. 1995. *HazCat Chemical Identification System Users Manual*. HazTech Systems, San Francisco, CA.

GLOSSARY

Accuracy Used to describe a monitor that is able to provide readings close to the actual amount of gas present.

Alpha Type of radioactive particle.

AP2C French chemical agent meter (CAM).

Advanced Portable Detection 2000 A warfare agent detection device that also monitors for mace, pepper spray, gamma radiation, and some toxic industrial chemicals.

APD2000 See *Advanced Portable Detection 2000.*

Beta Type of radioactive particle.

BID See *Biological indicating device.*

Boiling point The point at which a liquid changes to a gas; the closer to the boiling point, the more vapors that are produced.

Biomimetic A type of CO sensor used in home detectors.

Biological indicating device A detection device for biological agents, such as anthrax.

Bump test A test that uses a quantity and type of gas to ensure that a monitor responds and alarms to the gases being tested for.

C2 Canadian military chemical agent detection kit.

Calibration Checking the response of the monitor against known quantities of a sample gas, and if the readings differ, then electronically adjusting the monitor to read the same as the test gas.

CAM A detection device for nerve and blister agents; there are American, British, and French versions.

Catalytic bead The most common type of LEL sensor.

CGI See *Combustible gas indicator.*

Chemical Abstract Service A service that lists chemical substances and issues them a unique registration number, much like a social security number. Also known as CAS.

Chemical classifyer A testing strip that includes five tests; a companion test strip is the wastewater strip.

Chip measurement system A colorimetric sampling system that uses a bar-coded chip to measure known gases.

CMS See *Chip measurement system.*

Colorimetric A form of detection that involves a color change when the detection device is exposed to a sample chemical.

Colorimetric tubes Glass tubes filled with a material that changes color when the intended gas passes through the material.

Combustible gas indicator A monitor designed to measure the relative flammability of gases and to determine the percent of the lower explosive limit. Also known as LEL monitor.

Concentration The amount of corrosive as compared to water in a corrosive substance.

Curie Measurement of radioactive activity level.

Dosimeter Device that measures the body's dose of radiation.

Electrochemical sensor A sensor having an electrolyte gel that reacts with the intended substance, causing a reading on the instrument.

FID See *Flame ionization detector.*

Fire point The temperature with an ignition source at which a liquid ignites and sustains burning.

Flame ionization detector A device that uses a hydrogen flame to ionize a gas sample; used for the detection of organic materials.

Flammable range The numeric range between the lower explosive limit and the upper explosive limit in which a vapor will burn.

Flash point The temperature of a liquid that will produce sufficient vapors to form an ignitable mixture with air when an ignition source is present above the liquid.

Gamma Form of radioactive energy.

Gas chromatograph A detection device that splits chemical compounds and identifies them by their retention and travel times.

GC See *Gas chromatograph.*

Geiger-Mueller Detection device type for low energy materials.

GID-3 A detection device for nerve and blister agents.

Gray Measurement of radioactivity, equivalent to a RAD.

HazCat™ kit A chemical identification kit that can be used for solids, liquids, and gases.

Hyberbaric chamber A pressurized chamber that provides large amounts of oxygen to treat inhalation injuries, diving injuries, and other medical conditions.

Hydrolysis material The breakdown products of a material.

ICAM See *Improved chemical agent meter.*

IDLH See *Immediately dangerous to life or health.*

Immediately dangerous to life or health Term used by OSHA to describe an exposure level in which employees are at serious risk for becoming unable to remove themselves from the affected atmosphere after 30 minutes exposure. Levels much above IDLH or exposures greater than 30 minutes can be fatal.

Improved chemical agent meter A CAM in a different style box.

IMS See *Ion mobility spectrometry.*

Inert A chemical that is not toxic but will displace oxygen.

Infrared sensor A type of LEL sensor that uses infrared light to detect flammable gases.

Interferences Gases that are picked up by the sensor but are not intended to be read by the sensor.

Ionizing radiation Radiation that has enough energy to break up chemical bonds and can create ions. Examples include X rays, gamma radiation, and beta particles.

Ionization chamber Detection-type device for high energy materials.

Ionization potential The ability of a chemical to be ionized, or have its electrical charges separated and measured. To be read by a PID, a chemical must have a lower IP than the lamp in the PID.

Ion mobility See *Ion mobility spectrometry.*

Ion mobility spectrometry A detection technology that measures the travel time of ionized gases down a specific path.

IP See *Ionization potential.*

Lag time The amount of time it takes for a monitor to respond once exposed to a gas.

LEL See *Lower explosive limit.*

LEL meter The best name for a meter that is used to detect flammable gases.

LEL sensor A sensor designed to look for flammable gases; can be of four designs: Wheatstone bridge, catalytic bead, metal oxide, and infrared.

Lower Explosive limit The least amount of flammable gas and air mixture in which there can be a fire or explosion.

M-8 paper A paper detection device like pH paper that detects liquid nerve, blister, and VX agent.

M-9 paper A detection device used for liquid nerve and blister agents; comes in a tape form.

M90 chemical agent system A detection device for chemical warfare agents, was typically used for perimeter monitoring. It has been replaced by the GID-3 monitor.

M256A1 Kit A detection kit for nerve, blister, and choking agent vapors.

Mass spectrometer Almost always coupled with a GC, the MS measures the weight of a given.

Metal oxide sensor A form of LEL sensor.

Microrem Measurement of radiation; normal background radiation is usually in microrems.

Millirem Higher amount (1,000 times) of radiation than microrem.

Molecular weight The weight of the molecule based on the periodic table, or the weight of a compound when the atomic weights of the various components are combined.

MOS See *Metal oxide sensor.*

MS See *Mass spectrometer.*

National Fire Protection Association A consensus group that issues standards related to fire, HAZMAT, and other life safety concerns.

National Institute of Occupational Safety and Health The research agency of OSHA that studies worker safety and health issues.

NFPA See *National Fire Protection Association.*

NIOSH See *National Institute of Occupational Safety and Health.*

Nonionizing radiation Radiation that does not have enough energy to create charged particles, such as radio waves, microwaves, infrared light, visible light, and ultraviolet light.

Nuclear detonation Device that detonates through nuclear fission; the explosive power is derived from a nuclear source.

Occupational Safety and Health Administration Government agency tasked with providing safety regulations for workers.

Oleum Concentrated sulfuric acid which has been saturated with sulfur trioxide.

OSHA See *Occupational Safety and Health Administration.*

Oxygen deficient An oxygen level below 19.5 percent.

Oxygen enriched An oxygen level above 23.5 percent.

PEL See *Permissible exposure limit.*

Permissible exposure limit An occupational exposure limit established for an eight-hour period by OSHA.

Photoionization detector A detector that measures organic materials in the air by ionizing the gas with an ultraviolet lamp.

pH paper Testing paper used to indicate the corrosiveness of a liquid.

PID See *Photoionization detector.*

Precision The ability of a detector to repeat the results for a known atmosphere.

RAD Radiation absorbed dose; a quantity of radiation.

Radiation pager Detection device that alerts in the presence of gamma and X rays.

Radiological dispersion device Explosive device that spreads a radioactive material. The explosive power is derived from a nonnuclear source, such as a pipe bomb, which is attached to a radioactive substance.

Reagent Chemical material (solid or liquid) that is changed when exposed to a chemical substance.

Recovery time The amount of time it takes for a detector to return to zero after exposure to a gas.

Relative gas density Term used by the *NIOSH Pocket Guide* that means vapor density; a comparison of the weight of a gas to the weight of the air.

Relative response How a monitor reacts to a given gas as compared to the gas the monitor was calibrated for.

REM Radiation equivalent in man; a method of measuring radiation dose.

Risk-based response A system for identifying the risk chemicals present even though their specific identity is unknown; a system that characterizes all chemicals into fire, corrosive, or toxic risks.

Roentgen Basic unit of measurement for radiation.

SAW See *Surface acoustic wave.*

Sick building A building suspected of causing irritation or health problems for the occupants.

Sievert Unit of measurement equivalent to REM.

Smart sensors Sensors that have a computer chip on them that allows the switching of a variety of sensors within an instrument.

Smart ticket Detection device for some biological agents, such as anthrax.

Surface acoustic wave A sensor technology used in detection devices, such as the SAW and HazMat CADS. After the sampled gas passes over the sensor it outputs an algorithm which is checked for a possible match.

Threshold limit value An occupational exposure limit for an eight-hour day issued by the American Conference of Governmental Industrial Hygienists (ACGIH).

TLV See *Threshold limit value.*

Tote A portable container that holds solids, liquids, and gases. Liquid totes hold 300 to 500 gallons of various products; an increasingly common method of chemical storage and shipping.

Toxic gas sensor Device for the detection of toxic gases; common toxic gas sensors are for CO, H_2S, ammonia, and chlorine.

UEL See *Upper explosive limit.*

Upper explosive limit Upper range of the flammable range, the most flammable gas that can be present mixed with air to have a fire or explosion.

Vapor density How a gas responds relative to air. Air is given a value of 1, and gases with a vapor density less than 1 will rise, while those with a VD of greater than 1 will sink.

Vapor pressure The force of vapors coming from a liquid at a given temperature.

VD See *Vapor density.*

Volatility The amount of vapors coming from a liquid.

Wastewater strip A strip that does some additional tests beyond the chemical classifyer; a testing strip that does 5 tests.

Wheatstone bridge A form of LEL sensor.

X ray A form of radiation much like light but that bombards a target with electrons which makes it very penetrating.

ACRONYMS

ACGIH American Conference of Governmental Industrial Hygienists
AD Absorbed dose
APD Advanced Portable Detector
ASHRAE American Society of Heating, Refrigeration, and Air-Conditioning Engineers
ATSDR Agency for Toxic Substances and Diseases Registration
BID Biological indicating device
BRI Building-related illness
BTEX Benzene, toluene, ethylbenzene, and xylenes
CAM Chemical agent meter
CAS Chemical Abstract Service
CERCLA Comprehensive environmental response compensation liability act
CGI Combustible gas indicator
CMS Chip measurement system
CPM Count per minute
CRA Community Research Associates
DE Dose equivalent
EIRNS Emergency Incident Reporting and Notification System
EMS Emergency medical service
EPA Environmental Protection Agency
ETG Environmental Technology Group
FID Flame ionization detector
GC Gas chromatography
HAZMAT Hazardous materials
HAZWOPER Hazardous Waste Operations and Emergency Response
HCL Hydrochloric acid
HF Hydrofluoric acid
HMRU Hazardous Materials Response Unit
HVAC Heating, ventilation, and air-conditioning
IC Incident commander
ICAM Improved Chemical Agent Meter
IDLH Immediately dangerous to life and health
IMS Ion mobility spectrometry
IP Ionization potential
LCD Liquid crystal display
LEL Lower explosive limit
LFL Lower flammability limit
LPG Liquid petroleum gas
MCS Multiple chemical sensitivity
MDE Maryland Department of Environment
MFRI Maryland Fire and Rescue Institute
MOS Metal oxide sensor
MS Mass spectrometer
MSDS Material safety data sheet
NFPA National Fire Protection Association
NIOSH National Institute of Occupational Safety and Health
NRC Nuclear Regulatory Commission
OVA Organic vapor analyzer
OSHA Occupational Safety and Health Administration
PCB Polychlorinated biphenyl
PEL Permissible exposure limit
PETN Pentaerythritol tetranitrate
PID Photoionization detector
PPE Personal protective equipment
PVC Polyvinyl chloride
RAD Radiation absorbed dose
RBR Risk-based response
RDD Radiological dispersion device
REM Roentgen equivalent in man
RgasD Relative gas density
SAW Surface acoustic wave
SB Sick building
SBS Sick building syndrome
SCBA Self-contained breathing apparatus
SI International System of Units
TLV Threshold limit value
TOG Turnout gear
UDMH Unsymmetrical dimethyl hydrazine
UEL Upper explosive limit
UFL Upper flammability limit
UST Underground storage tank
UV Ultraviolet
VD Vapor density
VOC Volatile organic compound
WHO World Health Organization

INDEX